Learning WebRTC

Develop interactive real-time communication
applications with WebRTC

Dan Ristic

BIRMINGHAM - MUMBAI

Learning WebRTC

First published: June 2015

Production reference: 1250615

Published by Packt Publishing Ltd.
Livery Place
35 Livery Street
Birmingham B3 2PB, UK.

ISBN 978-1-78398-366-7

www.packtpub.com

Credits

Author
Dan Ristic

Reviewers
Roy Binux
Tsahi Levent-Levi
Andrii Sergiienko

Commissioning Editor
Andrew Duckworth

Acquisition Editor
Nikhil Karkal

Content Development Editor
Manasi Pandire

Technical Editor
Rahul C. Shah

Copy Editors
Sonia Michelle Cheema
Swati Priya
Neha Vyas

Project Coordinator
Bijal Patel

Proofreader
Safis Editing

Indexer
Monica Ajmera Mehta

Graphics
Disha Haria

Production Coordinator
Conidon Miranda

Cover Work
Conidon Miranda

About the Author

Dan Ristic is a frontend engineer and evangelist for Open Web. He strives to push the Web forward with new and creative ideas using the latest technologies. He attended the University of Advancing Technology, Arizona, where he furthered his engineering knowledge and fueled his passion for the Web. He has been writing applications on the Web ever since.

He currently lives and works in San Francisco as a senior software engineer at Sony Network Entertainment International. At Sony, he is responsible for managing the frontend application architecture and delivering the PlayStation Store to millions of users. In his free time, he can be found hiking, exploring, working on projects, and attending events.

I would like to thank my family for their never-ending support, motivation, and encouragement in everything I do.

I would also like to thank my friends for the innumerable cups of coffee and their help to keep me on track.

About the Reviewers

Roy Binux is a software engineer and an open source developer. He focuses on information extraction and the Spider System at work and builds interesting things online for fun. He is open to new technologies and tries to make the process of getting knowledge from the Web easier and build something useful with the power of the Web. Most of his work is open source, and you can find him on GitHub at `http://github.com/binux`.

Tsahi Levent-Levi is an independent analyst and consultant for WebRTC.

He has 15 years of experience in telecommunications, VoIP, and the 3G industry as an engineer, manager, marketer, and CTO. He is an entrepreneur, independent analyst, and consultant and assists companies to form a bridge between technologies and business strategies in the domain of telecommunications.

He has an MSc degree in computer science and an MBA degree, with specialization in entrepreneurship and strategy. He has been granted three patents related to 3G-324M and VoIP. He was the chairman of various activity groups within IMTC, an organization focusing on the interoperability of multimedia communications.

He is also the author and editor of his blog `https://bloggeek.me`, which focuses on the ecosystem and business opportunities around WebRTC.

Andrii Sergiienko is an entrepreneur who is passionate about information technology and travelling. He has lived in different places, such as Ukraine, Russia, Belarus, Mongolia, Buryatia, and Siberia—and has spent many years in every place. He also likes to travel by auto.

From his early childhood, he was interested in computer programming and hardware. He took his first step in these fields more than 20 years ago. He has experience working with a wide set of languages and technologies, including C, C++, Java, Assembler, Erlang, JavaScript, PHP, Riak, shell scripting, computer networks, security, and so on.

During his career, he has worked for both small local companies, such as domestic ISP, and large world corporations, such as Hewlett Packard. He has also started his own companies—some of them were relatively successful; others totally failed.

Today, he is working on the growth aspect of Oslikas—his new company that has its headquarters in Estonia. The company is focused on modern IT technologies and solutions. They also develop a full-stack framework to create rich media WebRTC applications and services. For more information on Oslikas, you can visit `http://www.oslikas.com`.

www.PacktPub.com

Support files, eBooks, discount offers, and more

For support files and downloads related to your book, please visit www.PacktPub.com.

Did you know that Packt offers eBook versions of every book published, with PDF and ePub files available? You can upgrade to the eBook version at www.PacktPub.com and as a print book customer, you are entitled to a discount on the eBook copy. Get in touch with us at service@packtpub.com for more details.

At www.PacktPub.com, you can also read a collection of free technical articles, sign up for a range of free newsletters and receive exclusive discounts and offers on Packt books and eBooks.

https://www2.packtpub.com/books/subscription/packtlib

Do you need instant solutions to your IT questions? PacktLib is Packt's online digital book library. Here, you can search, access, and read Packt's entire library of books.

Why subscribe?

- Fully searchable across every book published by Packt
- Copy and paste, print, and bookmark content
- On demand and accessible via a web browser

Free access for Packt account holders

If you have an account with Packt at www.PacktPub.com, you can use this to access PacktLib today and view 9 entirely free books. Simply use your login credentials for immediate access.

Table of Contents

Preface

When I first started writing HTML code, I was excited. Here I was typing letters into the keyboard, each one giving the computer some instructions that it knew what to do with. I was excited that I was creating something and that these instructions allowed me to express my creativity. When I finally saved my work, fired up my browser, and loaded my page, I was in awe. I saw my name in big bold letters with a moving picture, called a GIF, of an animated fire.

The Web has certainly come a long way since then. This is largely due to the fact that it is not just a place to create something new, but also a platform to share and show this creativity to others. This powerful platform for creative expression is what powers the Web and keeps it growing faster than ever. It has become so popular that we are moving our entire lives onto it. Websites power your e-mail, entertainment, bank accounts, legal documents, taxes, and even parts of this book were written using web tools. It is the want to move our lives to the Web that drives the development of powerful, easy-to-use APIs, such as WebRTC.

WebRTC is one of the most substantial additions to the Web platform. It brings about an entire suite of new technologies, such as cameras, streaming data, and even an entirely new network protocol stack. It is amazing to not only see the amount of work going into the WebRTC API, but also to know that this is all free for use by any application developer out there.

The aim of WebRTC is to democratize real-time communication. Earlier, building even a smaller video communication application used to take months and involved custom engineering to make even the smallest of applications. However, now we can do it in half the time or even less. This also brings the open source community into real-time communication. You can find other examples of WebRTC in the world and look at how these applications are built by searching through the source code.

It is this creative expression and freedom provided by the Web that drives the motivations for this book. I am glad to have the chance to bring it to more people through my writing, and I hope to inspire others just as I was inspired the first time I built a web page. Writing this book is one of the largest and toughest things I have done in my career, but I am grateful for all the help I have had along the way.

If you are looking for the easiest way to create a new real-time experience and share this with others, then read this book. This book, along with all technical books, is just a way of continuing to drive people to create something even better on the Web. You will learn not just about how to use WebRTC, but also what powers it under the hood. This book serves as not just a learning tool, but also as an inspiration for creating something truly amazing.

What this book covers

Chapter 1, Getting Started with WebRTC, covers how WebRTC enables audio and video communication for web-based applications. You will also begin by running an example of a WebRTC application inside your browser.

Chapter 2, Getting the User's Media, covers the first step when creating a communication application to get webcam and microphone input. This chapter also covers how to use the Media Capture and Streams API to capture this information from you. We also begin development by building the foundation of our communication example.

Chapter 3, Creating a Basic WebRTC Application, covers an introduction to the first WebRTC API — the RTCPeerConnection. This chapter also lays the groundwork for creating a WebRTC application by peeking inside the complex structure of WebRTC and what we can expect when we begin working with the API.

Chapter 4, Creating a Signaling Server, covers the steps in creating our very own signaling server to help our clients find each other on the Internet. This includes in-depth information on how signaling works in WebRTC and how we will utilize it in our example application.

Chapter 5, Connecting Clients Together, covers the actual usage of our signaling server. It also covers connecting two users successfully using the WebRTC API, Media Capture, and the signaling server that we created in the previous chapter to build our working example.

Chapter 6, Sending Data with WebRTC, covers an introduction to the RTCDataChannel and how it is used to send raw data between two peers. This chapter elaborates on our example by adding a text-based chat for our clients.

Chapter 7, File Sharing, elaborates on the concept of sending raw data by looking at how we can share files between two peers. This will demonstrate the many uses of WebRTC outside of audio and video sharing.

Chapter 8, Advanced Security and Large-scale Optimization, covers advanced topics when delivering a large-scale WebRTC application. We look at theoretical security and performance optimizations used by other companies in the industry.

Appendix, Answers to Self-test Questions, covers the answers to all the self-test questions that appear at the end of every chapter.

What you need for this book

All the examples in this book are built on the Web using web standards. Since the WebRTC specification is fairly new, it is recommended to run the examples inside an updated browser. It is preferred to use the latest Firefox or Chrome web browser for these.

All of the server code is written using Node.js. The Node.js framework runs inside most Windows, Linux, and Mac OSX machines.

You can use any text editor that supports the writing of JavaScript and HTML code.

Who this book is for

You should have some experience building web applications using HTML and JavaScript. Maybe, you are building an application right now, or have an idea of a new application that utilizes the power of audio and video communication between users. You might also need to deliver an application with high-performance data transfer between users.

You should have a firm grasp of programming concepts and web development, but the book is written for even a novice web engineer. The concepts are covered in-depth and taken one bit at a time, rather than charging ahead into the advanced topics. You may not know anything about WebRTC or have only heard a bit about it, and want to learn the inner workings of real-time communication.

Conventions

In this book, you will find a number of text styles that distinguish between different kinds of information. Here are some examples of these styles and an explanation of their meaning.

Code words in text, database table names, folder names, filenames, file extensions, pathnames, dummy URLs, user input, and Twitter handles are shown as follows: "The one thing to notice is that when we get an ICE candidate from `theirConnection`, we are adding it to our connection, and vice versa."

A block of code is set as follows:

```
<!DOCTYPE html>
<html lang="en">
  <head>
    <meta charset="utf-8" />
    <title>Learning WebRTC - Chapter 4: Creating a
        RTCPeerConnection</title>
  </head>
  <body>
    <div id=""container"">
      <video id=""yours"" autoplay></video>
      <video id=""theirs"" autoplay></video>
    </div>
    <script src=""main.js""></script>
  </body>
</html>
```

Any command-line input or output is written as follows:

```
> 1 + 1
2
> var hello = "world";
undefined
> "Hello" + hello;
'Helloworld'
```

New terms and **important words** are shown in bold. Words that you see on the screen, for example, in menus or dialog boxes, appear in the text like this: "Now, you should be able to click on the **Capture** button and capture one frame of the video feed on the canvas."

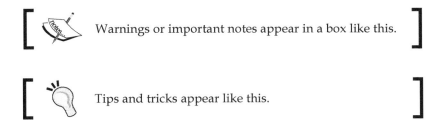

Warnings or important notes appear in a box like this.

Tips and tricks appear like this.

Reader feedback

Feedback from our readers is always welcome. Let us know what you think about this book—what you liked or disliked. Reader feedback is important for us as it helps us develop titles that you will really get the most out of.

To send us general feedback, simply e-mail feedback@packtpub.com, and mention the book's title in the subject of your message.

If there is a topic that you have expertise in and you are interested in either writing or contributing to a book, see our author guide at www.packtpub.com/authors.

Customer support

Now that you are the proud owner of a Packt book, we have a number of things to help you to get the most from your purchase.

Downloading the example code

You can download the example code files from your account at http://www.packtpub.com for all the Packt Publishing books you have purchased. If you purchased this book elsewhere, you can visit http://www.packtpub.com/support and register to have the files e-mailed directly to you.

Errata

Although we have taken every care to ensure the accuracy of our content, mistakes do happen. If you find a mistake in one of our books—maybe a mistake in the text or the code—we would be grateful if you could report this to us. By doing so, you can save other readers from frustration and help us improve subsequent versions of this book. If you find any errata, please report them by visiting `http://www.packtpub.com/submit-errata`, selecting your book, clicking on the **Errata Submission Form** link, and entering the details of your errata. Once your errata are verified, your submission will be accepted and the errata will be uploaded to our website or added to any list of existing errata under the Errata section of that title.

To view the previously submitted errata, go to `https://www.packtpub.com/books/content/support` and enter the name of the book in the search field. The required information will appear under the **Errata** section.

Piracy

Piracy of copyrighted material on the Internet is an ongoing problem across all media. At Packt, we take the protection of our copyright and licenses very seriously. If you come across any illegal copies of our works in any form on the Internet, please provide us with the location address or website name immediately so that we can pursue a remedy.

Please contact us at `copyright@packtpub.com` with a link to the suspected pirated material.

We appreciate your help in protecting our authors and our ability to bring you valuable content.

Questions

If you have a problem with any aspect of this book, you can contact us at `questions@packtpub.com`, and we will do our best to address the problem.

Getting Started with WebRTC

The Internet is no stranger to audio and video. Everyday web applications, such as Netflix and Pandora, stream audio and video content to millions of people. On the other hand, the Web is a stranger to real-time communication. Websites, such as Facebook, are only just starting to enable video-based communication in a browser, and they typically use a plugin that users have to install. This is where **Web Real-Time Communication (WebRTC)** comes into play.

In this chapter, we are going to cover the basics of WebRTC:

- The current status of the audio and video space
- The role that WebRTC plays in changing this space
- The major features of WebRTC and how they can be used

Audio and video communication today

Communicating with audio and video is a fairly common task with a history of technologies and tools. For a good example of audio communication, just take a look at a cell phone carrier. Large phone companies have established large networks of audio communication technology to bring audio communication to millions of people across the globe. These networks are a great example when it comes to showing widespread audio communication at its finest.

Video communication is also becoming just as prevalent as audio communication. With technologies such as Apple's FaceTime, Google Hangouts, and Skype video calling, speaking to someone over a video stream is a simple task for an everyday user. A wide range of techniques have been developed in these applications to ensure that the quality of the video is an excellent experience for the user. There have been engineering solutions to problems, such as losing data packets, recovering from disconnections, and reacting to changes in a user's network.

The aim of WebRTC is to bring all of this technology into the browser. Many of these solutions require users to install plugins or applications on their PCs and mobile devices. They also require developers to pay for licensing, creating a huge barrier and deterring new companies to join this space. With WebRTC, the focus is on enabling this technology for every browser user without the need for plugins or hefty technology license fees for developers. The idea is to be able to simply open up a website and connect with another user right then and there.

Enabling audio and video on the Web

The biggest accomplishment of WebRTC is bringing high-quality audio and video to the open the Web without the need for third-party software or plugins. Currently, there are no high-quality, well-built, freely available solutions that enable real-time communication in the browser. The success of the Internet is largely due to the high availability and open use of technologies, such as HTML, HTTP, and TCP/IP. To move the Internet forward, we want to continue building on top of these technologies. This is where WebRTC comes into play.

To build a real-time communication application from scratch, we would need to bring in a wealth of libraries and frameworks to deal with the many issues faced when developing these types of applications. These typically include software to handle connection dropping, data loss, and NAT traversal. The great thing about WebRTC is that all of this comes built-in to the browser API. Google has open sourced much of the technology involved in accomplishing this communication in a high-quality and complete manner.

 Most of the information about WebRTC, including the source code of its implementation, can be found freely available at http://www.webrtc.org/.

With WebRTC, the heavy lifting is all done for you. The API brings a host of technologies into the browser to make implementation details easy. This includes **camera and microphone capture, video and audio encoding and decoding, transportation layers**, and **session management**.

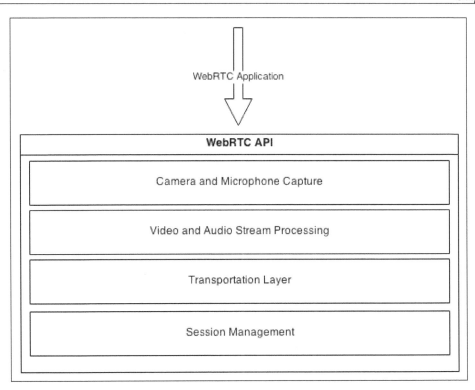

Camera and microphone capture

The first step to using any communication platform is to gain access to the camera and microphone on the device that the user is using. This means detecting the types of devices available, getting permission from the user to access them, and obtaining a stream of data from the device itself. This is where we will begin implementing our first application.

Encoding and decoding audio and video

Unfortunately, even with the improvements made in network speed, sending a stream of audio and video data over the Internet is too much to handle. This is where encoding and decoding comes in. This is the process of breaking down video frames or audio waves into smaller chunks and compressing them into a smaller size. The smaller size then makes it faster to send them across a network and decompress them on the other side. The algorithm behind this technique is typically called a **codec**.

If you have ever had trouble playing a video file on your computer, then you have some insight into the complex world of video and audio codecs. There are several different ways to encode audio and video streams, each with their different benefits. To add to this, there are many different companies that have different business goals behind creating and maintaining a codec. This means not all of the codecs are free for everyone to use.

There are many codecs in use inside WebRTC. These include H.264, Opus, iSAC, and VP8. When two browsers speak to each other, they pick the most optimal supported codec between the two users. The browser vendors also meet regularly to decide which codecs should be supported in order for the technology to work. You can read more about the support for various codecs at http://www.webrtc.org/faq.

You could easily write several books on the subject of codecs. In fact, there are many books already written on the subject. Fortunately for us, WebRTC does most of the encoding in the browser layer. We will not worry about it over the course of this book but, once you start venturing past basic video and audio communication, you will more than likely bump heads with codec support.

Transportation layer

The transportation layer is the topic of several other books as well. This layer deals with packet loss, ordering of packets, and connecting to other users. The API makes it easy to deal with the fluctuations of a user's network and facilitates reacting to changes in connectivity.

The way WebRTC handles packet transport is very similar to how the browser handles other transport layers, such as AJAX or WebSockets. The browser gives an easy-to-access API with events that tell you when there are issues with the connection. In reality, the code to handle a simple WebRTC call could be thousands or tens of thousands of lines long. These can be used to handle all the different use cases, ranging from mobile devices, desktops, and more.

Session management

Session management is the final piece of the WebRTC puzzle. This is simpler than managing network connectivity or dealing with codecs but still an important piece of the puzzle. This will deal with opening multiple connections in a browser, managing open connections, and organizing what goes to which person. This can most commonly be called **signaling** and will be dealt with more in *Chapter 4, Creating a Signaling Server*.

Included in this array of new features is also support for data transfer. Since a high-quality data connection is needed between two clients for audio and video, it also makes sense to use this connection to transfer arbitrary data. This is exposed to the JavaScript layer through the **RTCDataChannel** API. We will cover this in more detail at a later point.

WebRTC today has many of the building blocks needed to build an extremely high-quality real-time communication experience. Google, Mozilla, Opera, and many others have invested a wealth of time and effort through some of their best video and audio engineers to bring this experience to the Web. WebRTC even has roots in the same technology used to bring **Voice over Internet Protocol** (**VoIP**) communication to users. It will change the future of how engineers think about building real-time communication applications.

Creating web standards

The great thing about the Web is that it moves so fast. New standards are changed or created everyday and it is always improving. Browsers have further improved on this concept by allowing updates to be downloaded and installed without the user ever knowing. This makes the web developer's job an easier one, but it does mean that you have to keep up with what is going on in the world of the Web, and this includes WebRTC.

The way these changes are implemented across browsers is through standardized bodies. These are groups of individuals who work through a common organization to democratize the changing of browser APIs. The two organizations that control the standards for WebRTC are the **World Wide Web Consortium** (**W3C**) and the **Internet Engineering Task Force** (**IETC**).

Unlike many other standardized organizations, the W3C allows much of its information to be freely available to the public. This means that anyone can go online and view information about the implementation details of an API. The one for WebRTC is located at http://www.w3.org/TR/webrtc/. This is one way to refer to and learn more about how WebRTC works.

Getting involved in these organizations is one way to not only keep up-to-date on the latest technologies, but also to help shape the future of the Web. Participating in these communities makes browsers the fastest growing development stack out there. If you would like to learn more, visit http://www.w3.org/participate/ to find different ways to participate in the discussions.

Browser support

Although the goal of WebRTC is to be ubiquitous for every user, this does not mean that every browser all the same features at the same time. Different browsers may choose to be ahead of the curve in certain areas, which makes some things work in one browser and not another. The current support for WebRTC in the browser space is shown in the following section.

 There are multiple websites that can tell you if your browser supports a specific technology, such as http://caniuse.com/rtcpeerconnection, that tells you which browsers support WebRTC.

Compatibility with Chrome, Firefox, and Opera

There are chances that the browser you currently use supports WebRTC. Chrome, Firefox, and Opera all support WebRTC out-of-the-box. This should work on all mainstream OSes, such as Windows, Mac, and Linux, as well. The browser vendors, such as Chrome and Firefox, have also been working together to fix interoperability issues so they can all communicate with each other easily.

Compatibility with Android OS

This is also the case for Chrome and Firefox on Android operating systems as well. WebRTC-based applications should work out-of-the-box and be able to interoperate with other browsers after Android version 4.0 (Ice Cream Sandwich). This is due to the code sharing notion between the desktop and mobile versions of both Chrome and Firefox.

Compatibilty with Apple

Apple has made little effort to enable WebRTC in either Safari or iOS. There are rumors of support but no official date on when support will come about. The one workaround that others have used for hybrid native/web-based iOS applications is to embed the WebRTC code directly into their applications and load a WebRTC application into a **WebView**.

Compatibility with Internet Explorer

Microsoft has not announced any plans to enable WebRTC in Internet Explorer. They have proposed an alternative solution to enable audio and video communication in the browser. This alternative was turned down in favor of WebRTC. Since then, Microsoft has been a silent partner in the development of the technology.

 Throughout the course of this book, it is recommended that you use Chrome for all the examples. Always keep a lookout for updates on browser support, however, as this is a constantly changing space!

Using WebRTC in your browser

Now that you know which browser to use, we will jump right in and try out WebRTC right now! Navigate your browser to the demo application available at `https://apprtc.appspot.com/`. If you use Chrome, Firefox, or Opera, you should see a drop-down notification that looks similar to this:

Click on **Allow** to start streaming your audio and video input to the web page. You might have to configure your microphone or web camera settings to get them to work. Once you allow browser access to your camera and microphone, you should see a video feed of yourself from your camera.

The page should generate a custom ID for your current session. You should see this reflected in the URL of the page, such as `https://apprtc.appspot.com/r/359323927`. Simply copy and paste this URL into another browser window, either on your own computer or another one, and load the web page. Now, if everything works correctly, you should see two video feeds—one from your first client and another from the second. It should start to make sense why WebRTC is a powerful solution. This is how easy WebRTC makes real-time communication in the browser.

Applications enabled by WebRTC

Under the hood, WebRTC enables a basic peer-to-peer connection between two browsers. This is the heart of everything that happens with WebRTC. It is the first truly peer-to-peer connection inside a browser. This also means that anything you can do with peer connections can be easily extended to WebRTC. Many applications today use peer-to-peer capabilities, such as file sharing, text chat, multiplayer gaming, and even currencies. There are already hundreds of great examples of these types of applications working right inside the browser.

Most of these applications have one thing in common — they need a low-latency, high-performance connection between two users. WebRTC makes use of low-level protocols to deliver high-speed performance that could not be achieved otherwise. This speeds up data flow across the network, enabling large amounts of data to be transferred in a short amount of time.

WebRTC also enables a secure connection between two users to enable a higher level of privacy between them. Traffic traveling across a peer connection will not only be encrypted, but will also take a direct route to the other user. This means that packets sent in different connections might take entirely different routes over the Internet. This gives anonymity to users of WebRTC applications that is otherwise hard to guarantee when connecting to an application server.

This is just a subset of the types of applications enabled by WebRTC. Since WebRTC is built on the foundations of JavaScript and the Web, it can benefit many existing applications today. After reading this book, you should have the knowledge you need to create innovative WebRTC applications on your own!

Self-test questions

Q1. The goal of WebRTC is to provide easy access to real-time communication with no plugins or licensing fees. True or false?

Q2. Which of the following is not a feature that the browser provides through WebRTC?

1. Camera and microphone capture
2. Video and audio stream processing
3. Accessing a contact list
4. Session management

Q3. Participating in the W3C and IETC is only for big corporations with lots of money. True or false?

Q4. Which of the following is not a type of application that could benefit from using WebRTC?

1. File sharing
2. Video communication
3. Multiplayer gaming
4. None of the above

Summary

In this chapter, we gave you a glimpse of the features and technology behind WebRTC. You should have a firm grasp of what WebRTC aims to achieve and how this affects web applications today. You should also have an idea of what types of applications can be built with WebRTC and you should have tried out WebRTC already in your browser.

There was a wealth of information in this chapter, though if you did not take it all in, do not worry. We will go back and cover many of the topics presented here in detail over the course of the book. Feel free to explore some of the resources already on the Web today to get an even better understanding of what WebRTC is all about.

Next, we will start exploring camera and microphone capture using the `getUserMedia` API.

Then, we will start building our WebRTC application to handle a full one on one video and audio call directly in the browser.

Later on, we will start exploring how to extend this to multiple users, add data transfer through text-based chat and file sharing, and learn about the best security practices that are in place when using WebRTC.

2
Getting the User's Media

Obtaining a live video and audio feed from a user's webcam and microphone is the first step to creating a communication platform on WebRTC. This has traditionally been done through browser plugins, but we will use the getUserMedia API to do this all in JavaScript.

In this chapter, we will cover the following topics:

- Getting access to media devices
- Constraining the media stream
- Handling multiple devices
- Modifying the stream data

Getting access to media devices

There has been a long history behind trying to get media devices to the browser screen. Many have struggled with various Flash-based or other plugin-based solutions that required you to download and install something in your browser to be able to capture the user's camera. This is why the W3C decided to finally create a group to bring this functionality into the browser. The latest browsers now give the JavaScript access to the getUserMedia API, also known as the MediaStream API.

This API has a few key points of functionality:

- It provides a stream object that represents a real-time media stream, either in the form of audio or video
- It handles the selection of input devices when there are multiple cameras or microphones connected to the computer
- It provides security through user preferences and permissions that ask the user before a web page can start fetching a stream from a computer's device

Before we move any further, we should set a few standards about our coding environment. First off, you should have a text editor that allows you to edit HTML and JavaScript. There are tons of ways to accomplish this and, if you have purchased this book, the chances are high that you have a preferred editor already.

The other requirement for working on the media APIs is having a server to host and serve the HTML and JavaScript files. Opening up the files directly by double-clicking them will not work for the code presented in this book. This is due to the permissions and security set forth by the browser that does not allow it to connect to cameras or microphones unless it is being served by an actual server.

Setting up a static server

Setting up a local web server is the first step in any web developer's tool belt. In conjunction with text editors, static web servers are also plentiful and vary from language to language. My personal favorite is using Node.js with `node-static`, a great and easy-to-use web server:

1. Visit the Node.js website at `http://nodejs.org/`. There should be a big **INSTALL** button on the home page that will help you with installing Node.js on your OS.

2. Once Node.js is installed on your system, you will also have the package manager for Node.js installed called **node package manager** (**npm**).

3. Open up a terminal or command line interface and type `npm install -g node-static` (you will, more than likely, need administrator privileges).

4. Now you can navigate to any directory that contains the HTML files you would like to host on the server.

5. Run the `static` command to start a static web server in this directory. You can navigate to `http://localhost:8080` to see your file in the browser!

You will also be able to make directories and serve as many HTML files as you would like from this server. You can also use this to serve the example files associated with this book.

 There are many alternatives to using `node-static`, but we will need to use npm later, so I recommend you to get familiar with its syntax now.

Now we can move on to creating our first project!

Before you get started, you should make sure you have a camera and microphone attached to your computer. Most computers have settings options that will let you test your camera and make sure everything is working!

Creating our first MediaStream page

Our first WebRTC-enabled page will be a simple one. It will show a single `<video>` element on the screen, ask to use the user's camera, and show a live video feed of the user right in the browser. The `video` tag is a powerful HTML5 feature in itself. It will not only allow us to see our video on the screen, but can also be used to play back a variety of video sources. We will start by creating a simple HTML page with a `video` element contained in the `body` tag. Create a file named `index.html` and type the following:

```html
<!DOCTYPE html>
<html lang="en">
  <head>
    <meta charset="utf-8" />
    <title>Learning WebRTC - Chapter 2: Get User Media</title>
  </head>
  <body>
    <video autoplay></video>
    <script src="main.js"></script>
  </body>
</html>
```

Keep in mind that WebRTC is strictly an HTML5 feature. This means that you will have to use an up-to-date browser that supports the HTML5 standards. This can be seen by the DOCTYPE tag in our code that tells the browser to enter into a standard mode, which is compliant with HTML5, if it can.

If you open this page, there is nothing exciting going on yet. It should be a blank white page, which tells you that it is looking for the `main.js` file. We can start by adding the `main.js` file and add the `getUserMedia` code to it:

```js
function hasUserMedia() {
   return !!(navigator.getUserMedia || navigator.webkitGetUserMedia
|| navigator.mozGetUserMedia || navigator.msGetUserMedia);
}
```

```
if (hasUserMedia()) {
  navigator.getUserMedia = navigator.getUserMedia ||
navigator.webkitGetUserMedia || navigator.mozGetUserMedia ||
navigator.msGetUserMedia;
  navigator.getUserMedia({ video: true, audio: true }, function
(stream) {
    var video = document.querySelector('video');
    video.src = window.URL.createObjectURL(stream);
  }, function (err) {});
} else {
  alert("Sorry, your browser does not support getUserMedia.");
}
```

Now, you should be able to refresh your page and see everything in action! First, you should see a similar permission popup that you saw in the previous example when you ran the WebRTC demo. If you select **Allow**, it should get access to your camera and display your face in the <video> element on the web page.

The first step to work with the new browser APIs is to deal with the browser prefixes. Most of the time, browsers like to be ahead of the curve and implement features before they become an official standard. When they do this, they tend to create a prefix which is similar to the name of the browser (that is, **WebKit** for Chrome or **Moz** for Firefox). This allows the code to know if the API is a standard one or not, and deals with it accordingly. Unfortunately, this also creates several different methods for accessing the API in different browsers. We overcome this by creating a function to test if any of these functions exist in the current browser and, if they do, we assign them all to one common function that we can use in the rest of the code.

The next thing we do is to access the `getUserMedia` function. This function looks for a set of parameters (to customize what the browser will do) and a callback function. The callback function should accept one parameter: where the stream is coming from and the media devices on the computer.

This object points to a media stream that the browser is holding onto for us. This is constantly capturing data from the camera and microphone, and waiting for the instructions from the web application to do something with it. We then get the `<video>` element on the screen and load this stream into that element using `window.URL.createObjectURL`. Since elements cannot accept the JavaScript objects as parameters, it needs some string to fetch the video stream from. This function takes the stream object and turns it into a local URL that it can get the stream data from.

Note that the `<video>` element contains the attribute `autoplay`. If you remove this attribute, the stream will not show up when you assign it to the element. This is because the video expects to be *played* before it starts processing bytes from a video stream.

You have now completed the first step to build a WebRTC application! Reflecting on the code for this page, you can now see that the `getUserMedia` API is doing a lot for you. Even just getting a `stream` object from the camera and feeding it into a video on the screen could be the subject of an entire C or C++ book!

Constraining the media stream

Now that we know how to get a stream from the browser, we will cover configuring this stream using the first parameter of the `getUserMedia` API. This parameter expects an object of keys and values telling the browser how to look for and process streams coming from the connected devices. The first options we will cover are simply turning on or off the video or audio streams:

```
navigator.getUserMedia({ video: false, audio: true }, function
(stream) {
  // Now our stream does not contain any video!
});
```

When you add this stream to the `<video>` element, it will now not show any video coming from the camera. You can also do the opposite and get just a video feed and no audio. This is great while developing a WebRTC application when you do not want to listen to yourself talk all day!

 Typically, you will be calling yourself a lot while developing theWebRTC applications. This creates the phenomenon known as **audio feedback** where you are being recorded through the microphone and also playing those sounds through the speakers, creating an endless echo or loud ringing sound. Turning the audio off temporarily helps fix this issue during the development!

Let's take a look at the following screenshot, which illustrates a drop-down popup stating that `http://localhost/8080` needs access to the microphone:

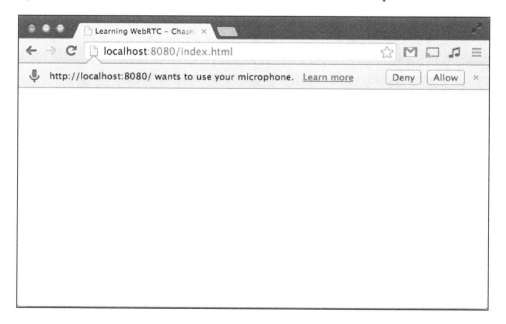

This can also be useful while creating a communication application that allows users to make audio-only calls, replicating a normal phone call. If someone does not want to share his/her video, it will only request access to the microphone in the browser. This will also prevent the camera from turning on or showing any lights, indicating that the browser is recording them.

Constraining the video capture

The options for constraining the `getUserMedia` API not only allows the `true` or `false` values, but also allows you to pass in an object with a complex set of constraints. You can see the full set of constraints provided in the specification detailed at `https://tools.ietf.org/html/draft-alvestrand-constraints-resolution-03`. These allow you to constrain options such as minimum required resolution and `frameRate`, video aspect ratio, and optional parameters all through the configuration object passed into the `getUserMedia` API.

This helps developers tackle several different scenarios that are faced while creating a WebRTC application. It gives the developer an option to request certain types of streams from the browser depending on the situation that the user is currently in. Some of these streams are listed here:

- Asking for a minimum resolution in order to create a good user experience for everyone participating in a video call
- Providing a certain width and height of a video in order to stay in line with a particular style or brand associated with the application
- Limiting the resolution of the video stream in order to save computational power or bandwidth if on a limited network connection

For instance, let's say that we wanted to ensure the video playback is always set to the aspect ratio of 16:9. This would be to avoid the video coming back in a smaller than desired aspect ratio, such as 4:3. If you change your `getUserMedia` call to the following, it will enforce the correct aspect ratio:

```
navigator.getUserMedia({
    video: {
      mandatory: {
        minAspectRatio: 1.777,
        maxAspectRatio: 1.778
      },
```

```
      optional: [
        { maxWidth: 640 },
        { maxHeigth: 480 }
      ]
    },
    audio: false
  }, function (stream) {
    var video = document.querySelector('video');
    video.src = window.URL.createObjectURL(stream);
  }, function (error) {
    console.log("Raised an error when capturing:", error);
  });
```

When you refresh your browser and give permission to the page to capture your camera, you should see the video is now wider than it used to be. In the first section of the configuration object, we gave it a mandatory aspect ratio of 16:9 or 1.777. In the optional section, we told the browser that we would like to stay under a width and height of 640 x 480. The optional block tells the browser to try and meet these requirements, if at all possible. You will probably end up with a 640 x 360 width and height for your video as this is a common solution to these constraints that most cameras support.

You will also notice that we passed in a second function to getUserMedia call. This is the error callback function and gets called if there are any issues with capturing the media stream. This could happen to you in the preceding example if your camera did not support 16:9 resolutions. Be sure to keep an eye on the development console in your browser to see any errors that get raised when this happens. If you successfully run this project, you can also change minAspectRatio or maxAspectRatio to see which parameters your browser can successfully run:

The power this gives us is the ability to adapt to the situations of the user's environment to provide the best video stream possible. This is incredibly helpful since the environment for browsers is vast and varied from user to user. If your WebRTC application plans to have a lot of users, you will have to find unique solutions to every unique environment. One of the biggest pains is supporting mobile devices. Not only do they have limited resources, but also limited screen space. You might want the phone to only capture a 480 x 320 resolution or smaller video stream in order to conserve power, processing, and bandwidth. A good way to test whether the user is on a mobile device is to use the user agent string in the browser and test it against the names of common mobile web browsers. Changing the getUserMedia call to the following will accomplish this:

```
var constraints = {
  video: {
    mandatory: {
      minWidth: 640,
      minHeight: 480
    }
  },
  audio: true
};
if (/Android|webOS|iPhone|iPad|iPod|BlackBerry|IEMobile|Opera
Mini/i.test(navigator.userAgent)) {
  // The user is using a mobile device, lower our minimum
resolution
  constraints = {
    video: {
      mandatory: {
        minWidth: 480,
        minHeight: 320,
        maxWidth: 1024,
        maxHeight: 768
      }
    },
    audio: true
  };
}
navigator.getUserMedia(constraints, function (stream) {
  var video = document.querySelector('video');
  video.src = window.URL.createObjectURL(stream);
}, function (error) {
  console.log("Raised an error when capturing:", error);
});
```

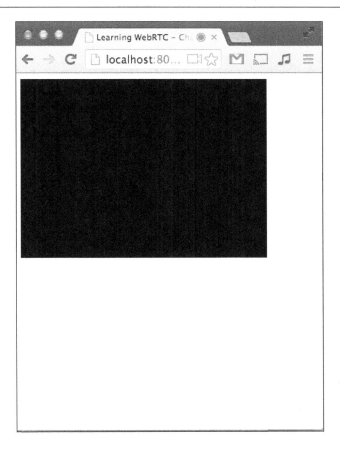

Constraints are not something to quickly glance over. They are the easiest way to increase the performance of a WebRTC application. While you are reading through the chapters in this book, you should be thinking about the different environments your application will run in and how you can best support each. We will also cover this in more depth in *Chapter 8, Advanced Security and Large-scale Optimization*, when we talk about WebRTC performance.

Handling multiple devices

In some cases, users may have more than one camera or microphone attached to their device. This is especially the case on mobile devices that often have a front-facing camera and a rear-facing one. In this case, you want to search through the available cameras or microphones and select the appropriate device for your user's needs. Fortunately, to do this, an API called MediaSourceTrack is exposed to the browser.

 On the other hand, since many of these APIs are still being created, not everything will be supported by all the browsers. This is especially the case with MediaSourceTrack, which is only supported in the latest version of Chrome at the time of writing this book.

With MediaSourceTrack, we can ask for a list of devices and select the one we need:

```
MediaStreamTrack.getSources(function(sources) {
    var audioSource = null;
    var videoSource = null;
    for (var i = 0; i < sources.length; ++i) {
      var source = sources[i];
      if(source.kind === "audio") {
        console.log("Microphone found:", source.label, source.id);
        audioSource = source.id;
      } else if (source.kind === "video") {
        console.log("Camera found:", source.label, source.id);
        videoSource = source.id;
      } else {
        console.log("Unknown source found:", source);
      }
    }
    var constraints = {
      audio: {
        optional: [{sourceId: audioSource}]
      },
      video: {
        optional: [{sourceId: videoSource}]
      }
    };
    navigator.getUserMedia(constraints, function (stream) {
      var video = document.querySelector('video');
      video.src = window.URL.createObjectURL(stream);
    }, function (error) {
      console.log("Raised an error when capturing:", error);
    });
});
```

This code calls getSources on MediaSourceTrack, which will give you a list of sources attached to the user's device. You can then iterate through them and select the one preferable to your application. If you open the development console while running this code, you will see the devices currently connected to the computer printed out. For instance, my computer has two microphones and one camera, as shown in the following screenshot:

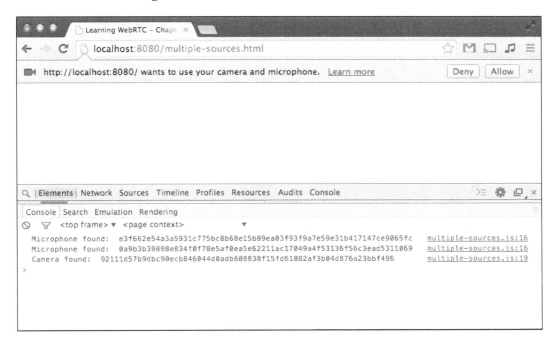

The source may also contain information such as which direction it faces to help with selection. With more progress and time, the browser could potentially provide a lot more information about the supported resolutions, **frames per second (fps)**, and more about the different devices available. Always be sure to research the latest updates on the getUserMedia and MediaStreamTrack API to see which browsers have added more features.

Creating a photo booth application

One of the best parts of the Web is that everything works together. This makes creating complex applications, such as a photo booth application, easy with other APIs like **Canvas**. A photo booth application allows you to see yourself on the screen while being able to capture pictures of yourself, much like a real photo booth. The Canvas API is a set of arbitrary methods to draw lines, shapes, and images on the screen. This is popularized through the use of Canvas for games and other interactive applications across the Web.

In this project, we are going to use the Canvas API to draw a frame of our video to the screen. It will take the current feed in our `video` element, translate it into a single image, and draw that image to a `<canvas>` element. We will set up our project with a simple HTML file:

```html
<!DOCTYPE html>
<html lang="en">
  <head>
    <meta charset="utf-8" />
    <title>Learning WebRTC - Chapter 2: Get User Media</title>
    <style>
      video, canvas {
        border: 1px solid gray;
        width: 480px;
        height: 320px;
      }
    </style>
  </head>
  <body>
    <video autoplay></video>
    <canvas></canvas>
    <button id="capture">Capture</button>
    <script src="photobooth.js"></script>
  </body>
</html>
```

Downloading the example code

You can download the example code files from your account at http://www.packtpub.com for all the Packt Publishing books you have purchased. If you purchased this book elsewhere, you can visit http://www.packtpub.com/support and register to have the files e-mailed directly to you.

All we have done is added a canvas to the page and are now looking for the photobooth.js file. Our JavaScript file is where the functionality lies:

```javascript
function hasUserMedia() {
  return !!(navigator.getUserMedia || navigator.webkitGetUserMedia
|| navigator.mozGetUserMedia || navigator.msGetUserMedia);
}
if (hasUserMedia()) {
  navigator.getUserMedia = navigator.getUserMedia ||
navigator.webkitGetUserMedia || navigator.mozGetUserMedia ||
navigator.msGetUserMedia;
  var video = document.querySelector('video'),
      canvas = document.querySelector('canvas'),
      streaming = false;
  navigator.getUserMedia({
    video: true,
    audio: false
  }, function (stream) {
    video.src = window.URL.createObjectURL(stream);
    streaming = true;
  }, function (error) {
    console.log("Raised an error when capturing:", error);
  });
  document.querySelector('#capture').addEventListener('click',
function (event) {
    if (streaming) {
      canvas.width = video.clientWidth;
      canvas.height = video.clientHeight;
      var context = canvas.getContext('2d');
      context.drawImage(video, 0, 0);
    }
  });
} else {
  alert("Sorry, your browser does not support getUserMedia.");
}
```

Now, you should be able to click on the **Capture** button and capture one frame of the video feed on the canvas. The image will show up as a single frame inside the `<canvas>` element. You can keep taking pictures to replace the image over and over again:

Modifying the media stream

We can take this project even further. Most image sharing applications today have some set of filters that you can apply to your images to make them look even cooler. This is also possible on the Web using CSS filters to provide different effects. We can add some CSS classes that apply different filters to our `<canvas>` element:

```
<style>
    .grayscale {
        -webkit-filter: grayscale(1);
        -moz-filter: grayscale(1);
        -ms-filter: grayscale(1);
        -o-filter: grayscale(1);
        filter: grayscale(1);
    }

    .sepia {
        -webkit-filter: sepia(1);
        -moz-filter: sepia(1);
        -ms-filter: sepia(1);
        -o-filter: sepia(1);
```

```
      filter: sepia(1);
    }

    .invert {
      -webkit-filter: invert(1);
      -moz-filter: invert(1);
      -ms-filter: invert(1);
      -o-filter: invert(1);
      filter: invert(1);
    }
  </style>
```

And, also some JavaScript to change the filter on click:

```
var filters = ['', 'grayscale', 'sepia', 'invert'],
    currentFilter = 0;
  document.querySelector('video').addEventListener('click',
function (event) {
    if (streaming) {
      canvas.width = video.clientWidth;
      canvas.height = video.clientHeight;

      var context = canvas.getContext('2d');
      context.drawImage(video, 0, 0);

      currentFilter++;
      if(currentFilter > filters.length - 1) currentFilter = 0;
      canvas.className = filters[currentFilter];
    }
  });
```

When you load this page, your snapshots should change whenever you take a new snapshot of the camera. This is utilizing the power of CSS filters to modify what the canvas is outputting. The browser then takes care of everything for you, such as applying the filter and showing the new image.

With the access to apply a stream to the canvas, you have unlimited possibilities. The canvas is a low-level and powerful drawing tool, which enables features such as drawing lines, shapes, and text. For instance, add the following after the class name is assigned to the canvas to add some text to your images:

```
context.fillStyle = "white";
context.fillText("Hello World!", 10, 10);
```

When you capture an image, you should see the text—**Hello World!**—placed in the upper-left corner of the image. Feel free to change the text, size, or more by changing the way the code uses the Canvas API. This can be taken even further using another Canvas technology called **WebGL**. This technology supports the 3D rendering right inside the browser and is quite an amazing accomplishment for JavaScript. You can utilize a streaming video source as a texture in WebGL and apply video to objects in the 3D space! There are thousands of interesting examples of this technology on the Web and I suggest you to look around a bit to see just what you can do in the browser.

Self-test questions

Q1. The browser will allow camera and microphone access even if the current page is opened as a file and not being served from a web server. True or false?

Q2. Which one of the following is not a correct browser prefix?

1. `webkitGetUserMedia`
2. `mozGetUserMedia`
3. `blinkGetUserMedia`
4. `msGetUserMedia`

Q3. The `getUserMedia` API will call the third argument as a function if an error happens while obtaining the camera or microphone stream. True or false?

Q4. Which one of the following does constraining the video stream not help with?

1. Securing the video stream
2. Saving processing power
3. Providing a good user experience
4. Saving bandwidth

Q5. The `getUserMedia` API can be combined with the Canvas API and CSS filters to add even more features to your application. True or false?

Summary

You should now have a firm grasp of the different ways to capture the microphone and camera streams. We also covered how to modify the constraints of the stream and select them from a set of devices. Our last project brought all of this together and added filters and image capture to create a unique photo booth application.

In this chapter, we covered how to get access to media devices, constrain the media stream, handle multiple devices, and modify the stream data.

The `MediaStream` specification is a constantly moving target. There are already plans to add a wealth of new features to the API enabling even more intriguing applications built on the Web. Always keep up with the latest specifications and support from the different browser vendors while developing WebRTC applications.

Although it may not seem like much, these techniques will help us in the upcoming chapters. Then knowing how to modify the input stream to ensure your WebRTC application is performing correctly will be the key.

In the upcoming chapters, we will use what we learned here to send our stream to another user using WebRTC.

3
Creating a Basic WebRTC Application

The first step of any WebRTC application is to create an RTCPeerConnection. Creating a successful RTCPeerConnection will require an understanding of the inner workings of how a browser creates peer connections. Firstly, in this chapter, we will lay the groundwork to understand the internals of WebRTC. Then we will utilize this knowledge to create a basic WebRTC video chat application.

In this chapter, we will cover the following topics:

- Understanding UDP transport and real-time transfer
- Signaling and negotiating with other users locally
- Finding other users on the Web and NAT traversal
- Creating an RTCPeerConnection

Understanding UDP transport and real-time transfer

Real-time transfer of data requires a fast connection speed between both the users. A typical connection needs to take a frame of both—audio and video—and send it to another user between 40 and 60 times per second in order to be considered good quality. Given this constraint, audio and video applications are allowed to miss certain frames of data in order to keep up the speed of the connection. This means that sending the most recent frame of data is more important than making sure that every frame gets to the other side.

A similar effect can already be seen with any video-playing application today. Video games and streaming media players can tolerate losing a few frames of video due to the special properties of the human brain. Our minds try to fill in the missing gaps as we visualize and process a video or game that we are watching. If our goal is to play 30 frames in one second and we miss frame 28, most of the time, the user will not even notice. This gives our video applications a different set of requirements:

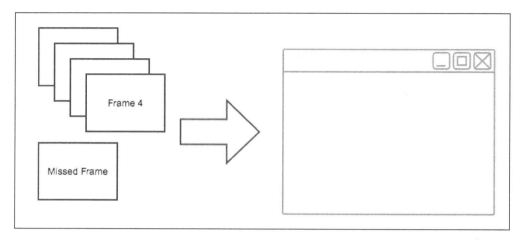

That is why **User Datagram Protocol (UDP)** is the transport protocol of choice when dealing with WebRTC applications. It gives us the power, or rather the lack of control, we need when dealing with a high-performance application. Most web applications today are built on top of the **Transmission Control Protocol (TCP)**. The reason for this is because of the guarantees it makes for its users, some of which are listed here :

- Any data sent will be acknowledged as received
- Any data that does not make it to the other side will get resent and halt the sending of any more data
- Data will be unique and no data will be duplicated on the other side

These features are the reason why TCP is a great choice for most things on the Web today. If you are sending an HTML page, it makes sense to have all the data come in the correct order with a guarantee that it got to the other side. Unfortunately, this technology is not a great fit for all use cases. Take, for instance, streaming data in a multiplayer game. Most data in a video game becomes stale in seconds or even less than that. This means that the user only cares about what has happened in the last few seconds and nothing more. If every piece of data needs to be guaranteed to make it to the other side, this can lead to a large bottleneck when the data goes missing:

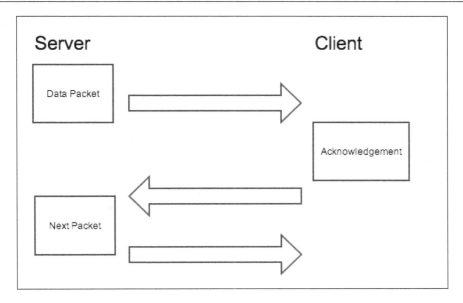

It is the need to work around the constraints of TCP that led the WebRTC developers to choose UDP as their preferred method of transport. The audio, video, and data requirements of WebRTC are not meant to be the most reliable connection, but rather to be the fastest one between the two browsers. This means we can afford to lose frames, which in turn means that UDP is a much better choice for these types of applications.

This does not mean that WebRTC never uses TCP as a mode of transportation. Later on, we will learn about **Traversal Using Relays around NAT** (**TURN**) servers and how they assist in transporting the WebRTC data between networks with heavy security using TCP.

UDP enables this scenario by making a lot of **non-guarantees**. It was built to be a less reliable transport layer that makes fewer assumptions about the data you are sending. You can see why in this list of things it does **not** guarantee:

- It does not guarantee the order your data is sent in or the order in which it will arrive on the other side

- It does not guarantee that every packet of data will make it to the other side; some may get lost along the way

- It does not track the state of every single data packet and will continue to send data even if data has been lost by the other client

Now, WebRTC can send audio and video in the fastest way possible. This should also reveal why WebRTC can be such a complex topic. Not every network allows UDP traffic across it. Large networks with corporate firewalls can block UDP traffic outright to try and protect against malicious connections. These connections have to travel along a different path than most of the web page downloads do today. Many workarounds and processes have to be built around UDP to get it to work properly for a wide audience. This is just the tip of the iceberg when it comes to WebRTC technology. In the next few sections, we will cover the other supporting technologies that enable WebRTC in the browser.

 UDP and TCP are not just used for web pages, but most Internet-based traffic you see today. You will find them being used in mobile devices, TVs, cars, and more. This is why it is important to understand these technologies, and how they work.

The WebRTC API

The next few sections will cover the WebRTC API currently implemented in the browser. These functions and objects allow developers to communicate with the WebRTC layer and make peer connections to other users. It consists of a few main pieces of technology:

- The RTCPeerConnection object
- Signaling and negotiation
- **Session Description Protocol (SDP)**
- **Interactive Connectivity Establishment (ICE)**

The RTCPeerConnection object

The RTCPeerConnection object is the main entry point to the WebRTC API. It is what allows us to initialize a connection, connect to peers, and attach media stream information. It handles the creation of a UDP connection with another user. It is time to get familiar with this name because you will be seeing it a lot throughout the rest of the book.

The job of the RTCPeerConnection object is to maintain the session and state of a peer connection in the browser. It also handles the setup and creation of a peer connection. It encapsulates all of these things and exposes a set of events that get fired at key points in the connection process. These events give you access to the configuration and internals of what happens during a peer connection:

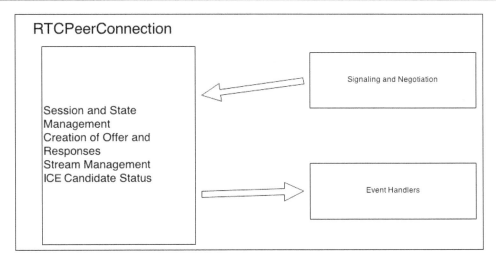

The RTCPeerConnection object is a simple object in the browser and can be instantiated using the new constructor as follows:

```
var myConnection = new RTCPeerConnection(configuration);
myConnection.onaddstream = function (stream) {
  // Use stream here
};
```

The connection accepts a configuration object, which we will cover later in this chapter. In the example, we have also added a handler for the onaddstream event. This is fired when the remote user adds a video or audio stream to their peer connection. We will also cover this later in the chapter.

Signaling and negotiation

Typically, connecting to another browser requires finding where that other browser is located on the Web. This is usually in the form of an IP address and port number, which act as a street address to navigate to your destination. The IP address of your computer or mobile device allows other Internet-enabled devices to send data directly between each other; this is what RTCPeerConnection is built on top of. Once these devices know how to find each other on the Internet, they also need to know how to talk to each other. This means exchanging data about which protocols each device supports as well as video and audio codecs and more.

This means that, in order to connect to another user, you need to know quite a bit about them. One possible solution would be to store a list on your computer of the users that you can connect to. To enable communication with another user, you would simply have to exchange contact information and let WebRTC handle the rest. This has the drawback, however, of your having to manually share information with each user that you want to connect to. You would have to maintain a big list of any users you wanted to connect with and exchange information through some other channel of communication. With WebRTC, we can make this process much more automated.

Luckily, the Web today has solved this problem in most communication applications we use today. To connect with anyone on popular services such as Facebook or LinkedIn, you just need to know their name and search for them. You can then add them to your list of known contacts and access their information at any time. This process is known as signaling and negotiation in WebRTC.

The process of signaling consists of a few steps:

1. Generate a list of potential candidates for a peer connection.
2. Either the user or a computer algorithm will select a user to make a connection with.
3. The signaling layer will notify that user that someone would like to connect with him/her, and he/she can accept or decline.
4. The first user is notified of the acceptance of the offer to connect.
5. If accepted, the first user will initiate `RTCPeerConnection` with the other user.
6. Both the users will exchange hardware and software information about their computers over the signaling channel.
7. Both the users will also exchange location information about their computers over the signaling channel.
8. The connection will either succeed or fail between the users.

This, however, is just an example of how WebRTC signaling *may* happen. In reality, the WebRTC specification does not contain any standards on how two users are supposed to exchange information. This is due to the ever-growing list of standards on connecting users. Many standards exist today, and even more are being created on the process of signaling and negotiating. The WebRTC standard writers decided that to try and agree on one standard would prevent it from moving forward.

In this book, we are going to create our own implementation of signaling and negotiation. This means writing a simple server that can transfer information between two browsers. Although it will be simple and prone to security flaws, it should give you a good understanding of how this process should work in WebRTC. At the same time, feel free to explore the numerous signaling options presented by many companies today. There are hundreds of signaling and negotiation solutions out there and more popping up every day. Some integrate with the current phone- or chat-based implementations, such as XMPP or SIP, and some come up with an entirely new way of signaling.

Session Description Protocol

To get connected with another user, you need to know a bit about them first. Some of the most important things to know about the other client is what audio and video codecs they support, how their network looks, and how much data their computer can handle. It also needs to be easily transportable between clients. Since we do not specify how this data should be transferred, it should also be capable of being sent over numerous types of transport protocols. This means we need a string-based business card with all the information about a user that we can send to other users. Luckily, this is exactly what SDP provides us with.

The great thing about SDP is that it has been around a long time, dating back to the late 90s for the first initial draft. This means that SDP is a tried-and-true method of establishing media-based connections between clients. It has been used in numerous other types of applications before WebRTC, such as phones and text-based chatting. This also means there are a lot of great resources out there on using and implementing it.

The SDP is a string-based data blob provided by the browser. The format of this string is a set of key-value pairs, all separated by line breaks:

```
<key>=<value>\n
```

The key is a single character that establishes the type of value this is. The value is a structured set of text that comprises a machine-readable configuration value. The different key-value pairs are then split by line breaks.

The SDP will cover the description, timing configuration, and media constraints for a given user. The SDP is given by the RTCPeerConnection object during the process of establishing a connection with another user. When we start working with the RTCPeerConnection object later in the chapter, you can easily print this to the JavaScript console. This will allow you to see exactly what is contained in the SDP, which may look something like this:

```
v=0
o=- 1167826560034916900 2 IN IP4 127.0.0.1
s=-
t=0 0
a=group:BUNDLE audio video
a=msid-semantic: WMS K44HTOZVjyAyAlvUVD3pOLu8i0LdytHiWRp1
m=audio 1 RTP/SAVPF 111 103 104 0 8 106 105 13 126
c=IN IP4 0.0.0.0
a=rtcp:1 IN IP4 0.0.0.0
a=ice-ufrag:Vl5FBUBecw/U3EzQ
a=ice-pwd:OtsNG6FzUH8uhNEhOg9/hprb
a=ice-options:google-ice
a=fingerprint:sha-256
FB:56:7D:B6:E0:C7:E7:39:FE:47:5A:12:6C:B4:4E:0E:2D:18:CE:AE:33:92:
A9:60:3F:14:E4:D9:AA:0D:BE:0D
a=setup:actpass
a=mid:audio
a=extmap:1 urn:ietf:params:rtp-hdrext:ssrc-audio-level
a=sendrecv
a=rtcp-mux
a=crypto:1 AES_CM_128_HMAC_SHA1_80
inline:zE+3pkUbJyFG4UmmvPxG/OFC4+QE24X8Zf3iOSCf
a=rtpmap:111 opus/48000/2
a=fmtp:111 minptime=10
a=rtpmap:103 ISAC/16000
a=rtpmap:104 ISAC/32000
a=rtpmap:0 PCMU/8000
a=rtpmap:8 PCMA/8000
a=rtpmap:106 CN/32000
a=rtpmap:105 CN/16000
a=rtpmap:13 CN/8000
a=rtpmap:126 telephone-event/8000
a=maxptime:60
a=ssrc:4274470304 cname:+j4Ma6UfMsCcQCWK
a=ssrc:4274470304 msid:K44HTOZVjyAyAlvUVD3pOLu8i0LdytHiWRp1
a1751f6b-98de-469b-b6c0-81f46e19009d
a=ssrc:4274470304 mslabel:K44HTOZVjyAyAlvUVD3pOLu8i0LdytHiWRp1
a=ssrc:4274470304 label:a1751f6b-98de-469b-b6c0-81f46e19009d
```

```
m=video 1 RTP/SAVPF 100 116 117
c=IN IP4 0.0.0.0
a=rtcp:1 IN IP4 0.0.0.0
a=ice-ufrag:Vl5FBUBecw/U3EzQ
a=ice-pwd:OtsNG6FzUH8uhNEhOg9/hprb
a=ice-options:google-ice
a=fingerprint:sha-256 FB:56:7D:B6:E0:C7:E7:39:FE:47:5A:12:6C:B4:4E:0E:
2D:18:CE:AE:33:92:
A9:60:3F:14:E4:D9:AA:0D:BE:0D
a=setup:actpass
a=mid:video
a=extmap:2 urn:ietf:params:rtp-hdrext:toffset
a=extmap:3 http://www.webrtc.org/experiments/rtp-hdrext/abs-send-
time
a=sendrecv
a=rtcp-mux
a=crypto:1 AES_CM_128_HMAC_SHA1_80
inline:zE+3pkUbJyFG4UmmvPxG/OFC4+QE24X8Zf3iOSCf
a=rtpmap:100 VP8/90000
a=rtcp-fb:100 ccm fir
a=rtcp-fb:100 nack
a=rtcp-fb:100 nack pli
a=rtcp-fb:100 goog-remb
a=rtpmap:116 red/90000
a=rtpmap:117 ulpfec/90000
a=ssrc:3285139021 cname:+j4Ma6UfMsCcQCWK
a=ssrc:3285139021 msid:K44HTOZVjyAyAlvUVD3pOLu8i0LdytHiWRp1
bd02b355-b8af-4b68-b82d-7b9cd03461cf
a=ssrc:3285139021 mslabel:K44HTOZVjyAyAlvUVD3pOLu8i0LdytHiWRp1
a=ssrc:3285139021 label:bd02b355-b8af-4b68-b82d-7b9cd03461cf
```

This is taken from my own machine during the session initiation process. As you can see, the code that is generated is complex to understand at first glance. It starts off by identifying the connection with the IP address. Then, it sets up basic information about the request such as whether I am requesting audio, video, or both. Next it sets up some audio information, including topics such as encryption type and the ice configuration. It also sets up the video information in the same manner. In the end, the goal is not to understand every line, but to get familiar with what the use of SDP is. You will never have to work with it directly during the course of this book, but may need to at some point in the future.

Overall, the SDP acts as a business card for your computer to other users trying to connect with you. The SDP, combined with signaling and negotiation, is the first half of the peer connection. In the next few sections, we will cover what happens after both users know how to find each other.

Finding a clear route to another user

A big part of most networks today is security. The chances are that any network you are using has several layers of access control, telling your data where and how it can be sent. This means that connecting to another user requires finding a clear path around not just your own network, but the other user's network as well. There are multiple technologies involved to achieve this inside WebRTC:

- **Session Traversal Utilities for NAT (STUN)**
- Traversal Using Relays around NAT (TURN)
- Interactive Connectivity Establishment (ICE)

These involve a number of servers and connections in order to be used properly by WebRTC. To understand how they work, we should first visualize how the layout of a typical WebRTC connection process looks like:

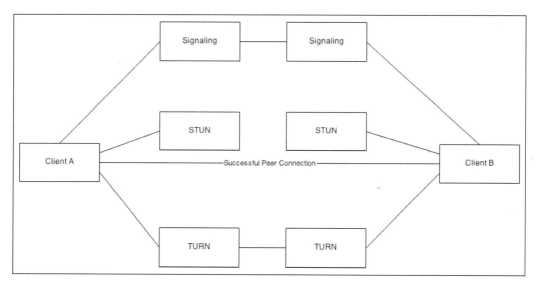

First off is finding out your IP address. Almost all devices connected to the Internet have an IP address, identifying their location on the Web. This is how you direct your data packets to the correct destination. The issue arises while finding your IP address in a network that is sitting behind a network router. The router hides your computer's IP address and replaces it with another one to increase security and allow multiple computers to use the same network address. Typically, you can have several IP addresses between yourself, your network router, and the public Internet.

Session Traversal Utilities for NAT

STUN is the first step in finding a good connection between two peers. It helps identify each user on the Internet, and is intended to be used by other protocols in making a peer connection. It starts by making a request to a server, enabled with the STUN protocol. The server then identifies the IP address of the client making the request, and returns that to the client. The client can then identify itself with the given IP address.

Using the STUN protocol requires having a STUN-enabled server to connect to. Currently, in Firefox and Chrome, default servers are provided directly from the browser vendors. This is great for getting up-and-running quickly and testing things out.

Although you may be praising the joys of serverless communication, setting up a good quality WebRTC application actually requires several servers to be enabled. You will need to provide your own set of STUN and TURN servers for your clients to use. There are plenty of great services already providing this today, so be sure to search around to find more information.

Traversal Using Relays around NAT

In some cases, a firewall might be too restrictive and not allow any STUN-based traffic to the other user. This may be the case in an enterprise NAT that utilizes port randomization to allow thousands of more devices than you would typically find. In this case, we need a different method of connecting with another user. The standard for this is called TURN.

The way this works is by adding a relay in between the clients that acts as a peer to peer connection on behalf of the client. The client then gets its information from the TURN server, much like streaming a video from a popular video service by making a request out to the server. This requires the TURN server to download, process, and redirect every packet that gets sent to it for each client. This is why, using TURN is often considered a last resort when making a WebRTC connection as the cost is high for setting up a quality TURN service.

There are many different statistics published on the use of STUN versus TURN, but they all seem to point to the same conclusion—most of the time, your users will be fine without TURN. The use of WebRTC with STUN will work with most network configurations. When setting up your own WebRTC service, it is a good idea to track this information and decide for yourself if the cost of using a TURN service is worth it.

You may notice that none of the examples have configuration values for TURN servers. The book assumes that the network you are on will be compatible with STUN. If you are having trouble connecting, it may be necessary to find a public low-use TURN server and use it while following the examples.

Interactive Connectivity Establishment

Now that we have covered STUN and TURN, we can learn how it is all brought together through another standard called ICE. It is the process that utilizes STUN and TURN to provide a successful route for peer to peer connections. It works by finding a range of addresses available to each user and testing each address in sorted order, until it finds a combination that will work for both the clients.

The process of ICE starts off by making no assumptions about each user's network configuration. It will incrementally go through a set of steps to discover how each client's network is set up. This process will use different sets of technologies to do this. The goal is to discover enough information about each network to make a successful connection.

Each ICE candidate is found through the use of STUN and TURN. It will query the STUN server to find the external IP address and append the location of a TURN server as a backup if the connection fails. Whenever the browser finds a new candidate, it notifies the client application that it needs to send the ICE candidate through the signaling channel. After enough addresses have been found and tested, and a connection is made, the process finally comes to an end.

Building a basic WebRTC application

Now that we have a good understanding of how the pieces of WebRTC are used, we are going to build our first WebRTC-enabled application. By the end of this chapter, you will have a working WebRTC web page where you can see the technology in action. We are going to pull all the information we just covered in to an easy-to-develop example. We will cover:

- Creating a RTCPeerConnection
- Creating the SDP `offer` and response
- Finding ICE candidates for peers
- Creating a successful WebRTC connection

Creating a RTCPeerConnection

The application we are going to create will, unfortunately, not be an entirely useful one, unless you happen to like looking at yourself in a mirror. What we aim to do in this chapter is connect a browser window to itself, streaming video data from the user's camera. The end goal is to get two video streams on the page, one coming from the camera directly and the other coming from a WebRTC connection that the browser has made locally.

Although this is not entirely helpful, it helps us by making the code much more readable and straightforward. We will learn how to use a server to make remote connection later on. Since the environment we are connecting to will be the same browser we are coming from, we do not have to worry about network instability or creating servers. After completing the project, your application should look something like this:

As you can see, other than my handsome face, this example is, otherwise, pretty basic. To start out, we will take similar steps to the first example we created in *Chapter 2, Getting the User's Media*. You will need to create another HTML page and host it using a local web server. It may be a good idea to refer to *Setting up a static server* subsection under *Getting access to media devices* section in *Chapter 2, Getting the User's Media*, and review how to set up your development environment.

The first step we will take is to create a few functions that handle support across multiple browsers. These will be able to tell us whether the current browser supports the functionality we need to use to make our application work. It will also normalize the API, making sure we can always use the same function, no matter what browser we may be running in.

Most browsers use prefixes on functions that are still in development. Be sure you are updated on the latest browser implementations to see what prefixes you should be using in your code. There are also the JavaScript libraries that help with dealing with prefixes.

To get started, set up a new web page with a JavaScript source file. Our HTML page will contain two video elements on it, one for the first client and another for the second:

```
<!DOCTYPE html>
<html lang="en">
  <head>
    <meta charset="utf-8" />
    <title>Learning WebRTC - Chapter 4: Creating a
        RTCPeerConnection</title>
  </head>
  <body>
    <div id=""container"">
      <video id=""yours"" autoplay></video>
      <video id=""theirs"" autoplay></video>
    </div>
    <script src=""main.js""></script>
  </body>
</html>
```

The html and head tags of this page should be familiar if you are used to creating HTML5 web pages. This is the standard format for any HTML5-compliant page. There are a lot of different boilerplate templates for creating a page, and this one is the one I feel is the simplest while still getting the job done. There is nothing that will drastically change the way our application works as long as the video elements are there, so if you need to make changes to this file, feel free to do so.

You will notice two video elements labeled — yours and theirs. These will be our two video feeds that will simulate connecting to another peer. Throughout the rest of this chapter — yours will be considered the local user that is initiating the connection. The other user — theirs — will be considered the remote user we are making the WebRTC connection to, even though they are not physically located somewhere else.

Lastly, we include our `script` function. Always keep in mind to add this at the end of the HTML page. This guarantees that the elements in the `body` are ready for use and the page is fully loaded for JavaScript to interact with it.

Next, we will create our JavaScript source code. Create a new file named `main.js` and start filling it out with the following code:

```
function hasUserMedia() {
   navigator.getUserMedia = navigator.getUserMedia ||
navigator.webkitGetUserMedia || navigator.mozGetUserMedia ||
navigator.msGetUserMedia;
   return !!navigator.getUserMedia;
}

function hasRTCPeerConnection() {
   window.RTCPeerConnection = window.RTCPeerConnection ||
window.webkitRTCPeerConnection || window.mozRTCPeerConnection;
   return !!window.RTCPeerConnection;
}
```

The first function deals with the `getUserMedia` API and should look familiar. The second function does a similar thing with the `RTCPeerConnection` object, ensuring we can use it in the browser. It first tries to assign any implemented WebRTC functions in the browser to a common function we can use in every use case. It then returns the assignment of that variable to see whether it actually exists in this browser.

Now that we can tell which APIs the user supports, let's go ahead and start using them. The next few steps should be pretty familiar as well. We are going to repeat some of the functionality we encountered in *Constraining the media stream* section in *Chapter 2, Getting the User's Media*, to get the user's camera stream. Before we do anything with WebRTC, we should get the local camera stream from the user. This ensures that the user is ready to create a peer connection and we do not have to wait for the user to accept camera sharing before we make a peer connection.

Most of the applications we build in WebRTC will go through a series of states. The hardest part about getting WebRTC to work is to do things in the proper order. If one step happens before another, it can break down the application quickly. These states are blocking, which means we cannot go onto the next state without completing the previous one. Here is an overview of how our application is going to work:

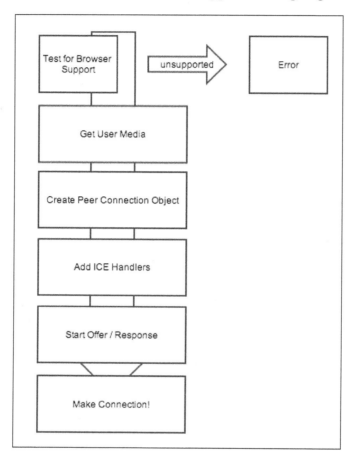

First we need to get the media stream from the user. This ensures that the stream is ready and the user has agreed to share their camera and microphone.

Next, we create the peer connection. This starts off the process in the disconnected state. This is where we can configure the WebRTC connection with the ICE servers that we would like to use. At this moment, the browser is sitting idly and waiting for the connection process to start.

The magic starts when an offer is created by one of the users. This kicks the browser into action and it starts to get ready to make a peer connection with another user. The offer and response are part of the signaling process, discussed in this chapter.

At the same time, the browser is also going to look for the candidate port and IP combinations that the other peer can connect to. It will continue to do this over a period of time, until a connection can be made or the connection fails to succeed. Once this is completed, the WebRTC connection process is over and the two users can start sharing information.

The next piece of code will capture the user's camera and make it available in our stream variable. You can now add this code after our two function definitions in your JavaScript:

```javascript
var yourVideo = document.querySelector(''#yours''),
    theirVideo = document.querySelector(''#theirs''),
    yourConnection, theirConnection;

if (hasUserMedia()) {
  navigator.getUserMedia({ video: true, audio: false }, function
(stream) {
    yourVideo.src = window.URL.createObjectURL(stream);

    if (hasRTCPeerConnection()) {
      startPeerConnection(stream);
    } else {
      alert(""Sorry, your browser does not support WebRTC."");
    }
  }, function (error) {
    alert(""Sorry, we failed to capture your camera, please try
again."");
  });
} else {
  alert("Sorry, your browser does not support WebRTC.");
}
```

The first part selects our video elements from the document and sets up a few variables we will use down the road. We are assuming that the browser supports the querySelector API at this point. We then check whether the user has access to the getUserMedia API. If they do not, our program stops here and we alert the user that they do not support WebRTC.

If this succeeds, we attempt to get the camera from the user. This is an asynchronous operation since the user has to agree to share their camera. If this succeeds, we set our local video's stream to the user's stream so that they can see this happened successfully. If this fails, we notify the user of the error and stop the process.

Finally, we check whether the user has access to the RTCPeerConnection API. If so, we call the function that will start our connection process (this will be defined in the next section). If not, we stop here and notify the user once more.

The next step is to implement the startPeerConnection function called in the previous section. This function will create our RTCPeerConnection objects, set up the SDP offer and response, and find the ICE candidates for both peers.

Now we create the RTCPeerConnection object for both the peers. Add the following to your JavaScript file:

```
function startPeerConnection(stream) {
  var configuration = {
    // Uncomment this code to add custom iceServers
    //"iceServers": [{ "url": "stun:stun.1.google.com:19302" }]"
}]
  };
  yourConnection = new webkitRTCPeerConnection(configuration);
  theirConnection = new webkitRTCPeerConnection(configuration);
};
```

Here, we define our function to create the connection objects. In the configuration object, you can pass parameters for which ICE servers you would like to use in your application. To use custom ICE servers, simply uncomment the code and change the value. The browser will automatically pick up the configuration and use it while making a peer connection. At this point, this is, however, not required, since the browser should have a default set of ICE servers to use. After this, we create two peer connection objects to represent each of the users in our application. Still keep in mind that both of our *users* will be in the same browser window for this application.

Creating the SDP offer and response answer

In this section, we will perform the offer and response answer process for making a peer connection. Our next block of code will set up the offer and response answer flows between the two peers:

```
function startPeerConnection(stream) {
  var configuration = {
```

```
    // Uncomment this code to add custom iceServers
    //""iceServers"": [{ ""url"": ""stun:stun.1.google.com:19302""
}]
  };
  yourConnection = new webkitRTCPeerConnection(configuration);
  theirConnection = new webkitRTCPeerConnection(configuration);

  // Begin the offer
  yourConnection.createOffer(function (offer) {
    yourConnection.setLocalDescription(offer);
    theirConnection.setRemoteDescription(offer);

    theirConnection.createAnswer(function (offer) {
      theirConnection.setLocalDescription(offer);
      yourConnection.setRemoteDescription(offer);
    });
  });
};
```

One thing you may notice is that, after an entire chapter's worth of explanation, this code seems rather simple. This is due to the fact that both peers are in the same browser window. This way, we can guarantee when the other user gets the offer and do not have to perform many asynchronous operations.

Implementing the offer/answer mechanism this way makes it easier to understand. You can clearly see the steps needed and the order they need to be in to successfully create a peer connection. If you are using a debugging tool attached to your browser, you can go through these steps and inspect the RTCPeerConnection objects at each step to see exactly what is happening.

In the next chapter, we will dig into this topic in a lot more depth. Typically, the other peer you will be connecting to will not be in the same browser — meaning a server is needed to connect peers between browser windows. This makes this process much more complex, given that these steps not only need to happen in the exact order they are shown here, but also across the multiple browser windows. This requires a lot of synchronization in an environment that may, sometimes, be unstable.

Finding ICE candidates

The last part of setting up the peer connection will be transferring the ICE candidates between the peers so that they can connect to each other. You can now change your `startPeerConnection` function to look like the following:

```
function startPeerConnection(stream) {
  var configuration = {
    // Uncomment this code to add custom iceServers
    //""iceServers"": [{ ""url"": ""stun:127.0.0.1:9876"" }]
  };
  yourConnection = new webkitRTCPeerConnection(configuration);
  theirConnection = new webkitRTCPeerConnection(configuration);

    // Setup ice handling
  yourConnection.onicecandidate = function (event) {
    if (event.candidate) {
      theirConnection.addIceCandidate(new
RTCIceCandidate(event.candidate));
    }
  };

    theirConnection.onicecandidate = function (event) {
    if (event.candidate) {
      yourConnection.addIceCandidate(new
RTCIceCandidate(event.candidate));
    }
    };

  // Begin the offer
  yourConnection.createOffer(function (offer) {
    yourConnection.setLocalDescription(offer);
    theirConnection.setRemoteDescription(offer);

    theirConnection.createAnswer(function (offer) {
      theirConnection.setLocalDescription(offer);
      yourConnection.setRemoteDescription(offer);
    });
  });
};
```

You may notice that this part of the code is completely event-driven. This is due to the asynchronous nature of finding the ICE candidates. The browser will continuously look for the candidates, until it has found as many as it thinks is good to create a peer connection or for the peer connection to be established and stable.

In the upcoming chapters, we will build out the functionality that actually sends this data across a signaling channel. The one thing to notice is that, when we get an ICE candidate from `theirConnection`, we are adding it to `yourConnection`, and vice versa. When we connect to someone who is not located in the same place that we are, this will have to travel across the Internet.

Adding streams and polishing

Adding a stream to a peer connection is easy with WebRTC. The API takes care of all the work of setting a stream up and sending its data across the wire. When the other user adds a stream to their peer connection, this notification gets sent across the connection, notifying the first user of the change. The browser then calls `onaddstream` to notify the user that a stream has been added:

```
// Setup stream listening
  yourConnection.addStream(stream);
  theirConnection.onaddstream = function (e) {
    theirVideo.src = window.URL.createObjectURL(e.stream);
  };
```

We can then add this stream to our local video by creating an object URL for the location of the stream. What this does is create a value that identifies the stream in the browser so that the `video` elements can interact with it. This acts as the unique ID for our video stream, telling the `video` element to play the video data coming from a local stream as the source.

Finally, we will add a little bit of styling to our application. The most popular style for a video communication application is one that is commonly seen in apps such as Skype. This has been replicated by many of the demos built with WebRTC today. Typically, the person you are calling is in the front and center of the application while your own camera is shown as a small window inside the larger one. Since we are building a web page, this can be achieved with some simple CSS as follows:

```
<style>
    body {
       background-color: #3D6DF2;
       margin-top: 15px;
    }
```

```
video {
  background: black;
  border: 1px solid gray;
}

#container {
  position: relative;
  display: block;
  margin: 0 auto;
  width: 500px;
  height: 500px;
}

#yours {
  width: 150px;
  height: 150px;
  position: absolute;
  top: 15px;
  right: 15px;
}

#theirs {
  width: 500px;
  height: 500px;
}
</style>
```

Simply add this to your HTML page, and you should have a good start to a well-styled WebRTC application. At this point, if you still think our applications looks dull, feel free to continue to add styling to the application. We will build on this in the upcoming chapters, and it is always more exciting to have a nice looking demo with some CSS.

Running your first WebRTC application

Now, run your web page to test it out. When you run the page it should ask you to share your camera with the browser. Once you accept, it will start the WebRTC connection process. The browser should almost instantly go through the steps that we have discussed so far, and create a connection. You should then see two videos of yourself, one from your camera and the other being streamed over a WebRTC connection.

For reference, here is a full listing of the code from this example. The following is the code from our `index.html` file:

```html
<!DOCTYPE html>
<html lang=""en"">
  <head>
    <meta charset=""utf-8"" />

    <title>Learning WebRTC - Chapter 4: Creating a
RTCPeerConnection</title>

    <style>
      body {
        background-color: #3D6DF2;
        margin-top: 15px;
      }

      video {
        background: black;
        border: 1px solid gray;
      }

      #container {
        position: relative;
        display: block;
        margin: 0 auto;
        width: 500px;
        height: 500px;
      }

      #yours {
        width: 150px;
        height: 150px;
        position: absolute;
        top: 15px;
        right: 15px;
      }

      #theirs {
        width: 500px;
        height: 500px;
      }
    </style>
  </head>
```

```html
<body>
  <div id=""container"">
    <video id=""yours"" autoplay></video>
    <video id=""theirs"" autoplay></video>
  </div>

  <script src=""main.js""></script>
</body>
</html>
```

The following is the code from our `main.js` JavaScript file:

```javascript
function hasUserMedia() {
  navigator.getUserMedia = navigator.getUserMedia ||
navigator.webkitGetUserMedia || navigator.mozGetUserMedia ||
navigator.msGetUserMedia;
  return !!navigator.getUserMedia;
}

function hasRTCPeerConnection() {
  window.RTCPeerConnection = window.RTCPeerConnection ||
window.webkitRTCPeerConnection || window.mozRTCPeerConnection;
  return !!window.RTCPeerConnection;
}

var yourVideo = document.querySelector(''#yours''),
    theirVideo = document.querySelector(''#theirs''),
    yourConnection, theirConnection;

if (hasUserMedia()) {
  navigator.getUserMedia({ video: true, audio: false }, function
(stream) {
    yourVideo.src = window.URL.createObjectURL(stream);

    if (hasRTCPeerConnection()) {
      startPeerConnection(stream);
    } else {
      alert(""Sorry, your browser does not support WebRTC."");
    }
  }, function (error) {
    console.log(error);
  });
```

```
} else {
  alert(""Sorry, your browser does not support WebRTC."");
}

function startPeerConnection(stream) {
  var configuration = {
    ""iceServers"": [{ ""url"": ""stun:stun.1.google.com:19302""
}]
  };
  yourConnection = new webkitRTCPeerConnection(configuration);
  theirConnection = new webkitRTCPeerConnection(configuration);

  // Setup stream listening
  yourConnection.addStream(stream);
  theirConnection.onaddstream = function (e) {
    theirVideo.src = window.URL.createObjectURL(e.stream);
  };

  // Setup ice handling
  yourConnection.onicecandidate = function (event) {
    if (event.candidate) {
      theirConnection.addIceCandidate(new RTCIceCandidate(event.
candidate));
    }
  };

  theirConnection.onicecandidate = function (event) {
    if (event.candidate) {
      yourConnection.addIceCandidate(new RTCIceCandidate(event.
candidate));
    }
  };

  // Begin the offer
  yourConnection.createOffer(function (offer) {
    yourConnection.setLocalDescription(offer);
    theirConnection.setRemoteDescription(offer);

    theirConnection.createAnswer(function (offer) {
      theirConnection.setLocalDescription(offer);
      yourConnection.setRemoteDescription(offer);
    });
  });
};
```

Self-test questions

Q1. UDP is well suited for WebRTC peer connections because of the non-guarantees it makes when delivering data packets. True or false?

Q2. Signaling and negotiation is part of the WebRTC standard and is completely taken care of by the browser. True or false?

Q3. The Session Description Protocol (SDP) is best described as:

1. A configuration file for WebRTC
2. A way to figure out what video codecs are supported
3. A business card for your computer
4. A confusing technical document that no one understands

Q4. Interactive Connectivity Establishment (ICE) assists in finding a clean path between two clients in a typical network setup. True or false?

Q5. Which is not true about TURN?

1. It requires more bandwidth and processing power than a normal connection
2. TURN should be a last resort after trying other methods of connection
3. A TURN server will have to process every packet sent between clients
4. TURN methods are provided by the browser

Summary

Congratulations on making it this far. If you have successfully completed this chapter, you are well on your way to making larger WebRTC applications. The goal of the this chapter was not only to create a WebRTC application but also to understand what happens at each step of the process.

After this chapter, it should already be clear that WebRTC is a complex piece of technology. We covered a great deal of information on the inner workings of WebRTC. Although it is not required to know every bit about how WebRTC is implemented in browsers today, having an understanding of how the major parts work together will help you understand the examples to come.

In this chapter, we covered the inner workings of how peer connections are created in the browser. We covered several of the technologies that enable this, including UDP, SDP, and ICE. You should now have a surface-level understanding of how two browsers can find each other and communicate over the Internet.

It would be a good idea to review the material we have covered so far to fully understand how WebRTC works in our example. It is important to note that each step is important as well as the sequence. This will help debug issues in your WebRTC application as we introduce more complexity in future chapters.

The rest of the book will build upon this example, making it much more complex than it is currently. We will add features to connect multiple users across several browsers in many different environments. Each chapter will take a part of the WebRTC process and look at it in depth, covering the common pitfalls, and take care of edge cases such as network stability and security.

In the next chapter, we will begin building the signaling server to support connecting remote users. This is the basis for the signaling server that we will use through the rest of the book. It will also allow us to create our first real calling application, much like Google Hangouts.

Creating a Signaling Server

4

At some point when creating a WebRTC application, you will have to break away from developing for a client and build a server. Most WebRTC applications are not solely dependent on just being able to communicate through audio and video and typically need many other features to be interesting. In this chapter, we are going to dive into server programming using JavaScript and Node.js. We are going to create the basis for a basic signaling server that we can utilize through the rest of the book.

In this chapter, we will cover the following topics:

- Setting up our environment to develop in Node.js
- Connecting to the client using WebSockets
- Identifying users
- Initiating and answering a WebRTC call
- Handling ICE candidate transfers
- Hanging up a call

Throughout this chapter, we are going to focus solely on the server part of the application. In the next chapter, we will build the client part of this example. Our example server will be bare bones in nature, giving us just enough to set up a WebRTC peer connection.

Building a signaling server

The server we are going to build in this chapter will help us connect two users together who are not located on the same computer. The goal of the server is to replace the signaling mechanism with something that travels over a network. The server will be straightforward and simple, supporting only the most basic WebRTC connections.

Our implementation will have to respond to and answer requests from multiple users. It will do this by having a simple bidirectional messaging system between clients. It will allow one user to call another and setup a WebRTC connection between them. Once a user has called another, the server will pass the offer, answer, and ICE candidates between the two users. This will allow them to successfully setup a WebRTC connection.

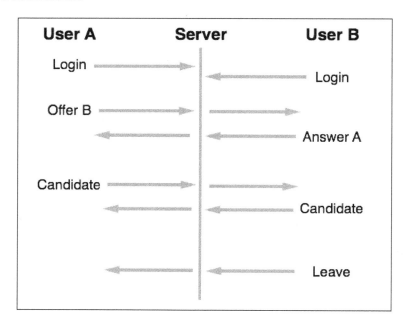

The preceding diagram shows the messaging flow between clients when using the signaling server to setup a connection. Each side will start by registering themselves with the server. Our logging in will simply send a string-based user identifier to the server and make sure it is not taken already. Once both users have registered with the server, they can then call another user. Making an offer with the user identifier they wish to call does this. The other user should answer in turn. Finally, candidates are sent between clients until they can successfully make a connection. At any point, a user can terminate the connection by sending the **leave** message. The implementation will be simple, acting mostly as a pass-through for the users to send messages to each other.

Keep in mind that this is just one example of a signaling server. Since there are no rules when implementing signaling, you can use any protocol, technology, or pattern that you like!

Setting up our environment

We are going to utilize the power of Node.js to build our server. If you have never programmed in Node.js before, do not worry! This technology utilizes a JavaScript engine to do all the work. This means that all of the programming will be in JavaScript so there will be no new language to learn. Now, let's perform the following steps to set up our Node.js environment:

1. The first step to running a node.js server is to install node.js.

 You can visit the website at `http://nodejs.org/` to get more information on how to install it. There are many options out there for every OS, so feel free to choose the one that best suits your needs. At the time of writing, the latest Node.js version is v0.12.4.

2. Now you can open up your terminal application and use the `node` command to bring up the Node.js VM. Node.js is based on the V8 JavaScript engine that comes with Google Chrome. This means that it works extremely close to how the browser interprets JavaScript. Type in a few commands to get used to how it works:

```
> 1 + 1
2
> var hello = "world";
undefined
> "Hello" + hello;
'Helloworld'
```

3. From here, we can start creating our server program. Luckily, Node.js runs JavaScript files as well as commands typed in the terminal. Create an `index.js` file with the following contents and run it using the `node index.js` command:

```
console.log("Hello from node!");
```

When you run the `node index.js` command, you will see the following output in the Node.js terminal:

```
Hello from node!
```

This is the end of the Node.js concepts that we will cover in this book. Our implementation of a signaling server is not the most advanced, and digging into server engineering would require an entire second book's worth of content. As we move on, take time to learn more about Node.js or even translate the signaling server that we will build into your favorite language!

Getting a connection

The steps required to create a WebRTC connection are required to be real-time. This means that clients will have to be able to transfer messages between each other in real time without using a WebRTC peer connection. This is where we will utilize another powerful feature of HTML5 called WebSockets.

A WebSocket is exactly what it sounds like—an open bidirectional socket connection between two endpoints—a web browser and a web server. You can send messages back and forth over the socket using strings and binary information. It is designed to be implemented in both web browsers and web servers to enable communication between them, outside of the realm of AJAX requests.

The WebSocket protocol has been around since about 2010 and is a well-defined standard that is available in most browsers today. It has wide support across web clients, and many server technologies have frameworks dedicated to their use. There are even entire frameworks that rely on WebSocket technology such as the **Meteor JavaScript framework**.

The big difference between the WebSocket protocol and the WebRTC protocol is the use of the TCP stack. WebSockets has been designed to be client-to-server in nature and utilizes TCP transport for a reliable connection. This means it has many of the bottlenecks that WebRTC does not have, which we described in *Understanding UDP transport and real-time transfer* section in *Chapter 3, Creating a Basic WebRTC Application*. This is also the reason that it works well as a signaling transport protocol. Since it is reliable, our signals are less likely to get dropped between users, giving us more successful connections. It is also built into the browser and makes it easy to set up using Node.js, which makes the implementation of our signaling server easier to understand.

To utilize the power of WebSockets in our project, we must first install a supported WebSockets library for Node.js. We will be using the ws project from the npm registry. To install the library, navigate to the directory of the server and run the following command:

```
npm install ws
```

You should see the following output:

```
dan:webrtc-book/ (master) $ npm install ws
npm http GET https://registry.npmjs.org/ws
npm http 304 https://registry.npmjs.org/ws
npm http GET https://registry.npmjs.org/tinycolor
npm http GET https://registry.npmjs.org/options
npm http GET https://registry.npmjs.org/commander
npm http GET https://registry.npmjs.org/nan
npm http 304 https://registry.npmjs.org/tinycolor
npm http 304 https://registry.npmjs.org/options
npm http 200 https://registry.npmjs.org/commander
npm http 200 https://registry.npmjs.org/nan

> ws@0.4.31 install /Users/dan/workspace/webrtc-book/node_modules/ws
> (node-gyp rebuild 2> builderror.log) || (exit 0)

  CXX(target) Release/obj.target/bufferutil/src/bufferutil.o
  SOLINK_MODULE(target) Release/bufferutil.node
  SOLINK_MODULE(target) Release/bufferutil.node: Finished
  CXX(target) Release/obj.target/validation/src/validation.o
  SOLINK_MODULE(target) Release/validation.node
  SOLINK_MODULE(target) Release/validation.node: Finished
ws@0.4.31 node_modules/ws
├── tinycolor@0.0.1
├── options@0.0.5
├── commander@0.6.1
└── nan@0.3.2
```

 npm is a package manager for Node.js. It hosts and keeps a list of open source frameworks that anyone can download and use in their projects. Navigate to https://www.npmjs.org/ for more information on this subject.

Now that we have installed the WebSocket library, we can start using it in our server. You can insert the following code in our index.js file:

```
var WebSocketServer = require(''ws'').Server,
    wss = new WebSocketServer({ port: 8888 });

wss.on(''connection'', function (connection) {
  console.log(""User connected"");

  connection.on(''message'', function (message) {
    console.log(""Got message:"", message);
  });

  connection.send(''Hello World'');
});
```

The first line requires the WebSocket library that we installed in our previous command. We then create the WebSocket server, telling it what port to connect to listen on. You can specify any port you would like if you need to change this setting.

Next, we listen to the connection event coming from the server. This code will get called whenever a user makes a WebSocket connection to the server. It will give you a connection object that has all sorts of information about the user who has just connected.

We then listen to any messages that are being sent by the user. For now, we just log these messages to the console.

Finally, we send a response to the client saying Hello World. This happens immediately when the server has completed the WebSocket connection with the client.

Note that the connection event happens for any user connecting to the server. This means that you can have multiple users connecting to the same server and each one will trigger the connection event individually. This asynchronous-based code is often seen as one of the strong points of programming in Node.js.

Now we can run our server by running node index.js. The process should start and simply wait to handle WebSocket connections. It will do this indefinitely until you stop the process from running.

Testing our server

To test whether our code is functioning properly, we can use the wscat command that comes with the ws library. The great thing about npm is that you cannot only install libraries to use in your application, but also install libraries globally to be used as command-line tools. The way to do this is by running npm install -g ws, although you might need to use administrator privileges when running this command.

This should give us a new command called wscat. This tool allows us to connect directly to WebSocket servers from the command line and test out commands against them. To do this, we run our server in one terminal window, then open a new one and run the wscat -c ws://localhost:8888 command. You will notice ws://, which is the custom directive for the WebSocket protocol instead of HTTP. Your output should look similar to this:

```
dan:webrtc-book/ (masterx) $ wscat -c ws://localhost:8888
connected (press CTRL+C to quit)
< Hello World
>
```

Your server should also log the connection to its console:

If either of these do not work, then check the code against the listing and read the documentation for the ws library as well as Node.js and npm. These tools may work differently in different environments and require extra setting up in some cases. If everything does work, pat yourself on the back for writing a WebSocket server in Node.js with 12 lines of code.

Identifying users

In a typical web application, the server will need a way to identify between connected clients. Most applications today use the one-identity rule and have each user login to a respective string-based identifier known as their username. We will also be using the same rule in our signaling application. It will not be as sophisticated as some of the methods used today since we will not even require a password from the user. We simply need an ID for each connection so we know where to send messages.

To start, we are going to change our connection handler a bit, to look similar to this:

```
connection.on(''message'', function (message) {
    var data;

    try {
      data = JSON.parse(message);
    } catch (e) {
      console.log(""Error parsing JSON"");
      data = {};
    }
});
```

This will change our WebSocket implementation to only accept JSON messages. Since WebSocket connections are limited to strings and binary data, we need a way to send structured data across the wire. JSON allows us to define structured data and then serialize it to a string that we can send over a WebSocket connection. It is also the easiest form of serialization to use in JavaScript.

Next, we will need a way to store all of our users who are connected. Since our server is simplistic in nature, we will use a hash-map otherwise known in JavaScript as an object, to store our data. We can change the top of our file to look similar to this:

```
var WebSocketServer = require(''ws'').Server,
    wss = new WebSocketServer({ port: 8888 }),
    users = {};
```

To login, we will need to know that a user is sending a `login` type message. To support this, we are going to add a `type` field to every message that is sent from the client. This will allow our server to know what to do with the data that it is receiving. Firstly, we will define what to do when the user tries to login:

```
connection.on(''message'', function (message) {
    var data;

    try {
        data = JSON.parse(message);
    } catch (e) {
        console.log(""Error parsing JSON"");
        data = {};
    }

    switch (data.type) {
        case ""login"":
            console.log(""User logged in as"", data.name);
            if (users[data.name]) {
                sendTo(connection, {
                    type: ""login"",
                    success: false
                });
            } else {
                users[data.name] = connection;
                connection.name = data.name;
                sendTo(connection, {
                    type: ""login"",
                    success: true
                });
            }

            break;
        default:
            sendTo(connection, {
```

```
        type: ""error"",
        message: ""Unrecognized command: "" + data.type
    });

    break;
  }
});
```

We use a `switch` statement to handle each message type accordingly. If the user sends a message with the `login` type, we first need to see if anyone has already logged into the server with that ID. If they have, we tell the client that they have not successfully logged in and need to pick a new name. If no one is using this ID, we set the connection to a key in our user's object with the ID being the key. If we run into any commands we do not recognize, we also send a message back to the client saying there was an error processing their request.

I also added a helper function called `sendTo` in the code that handles sending a message to a connection. This can be added anywhere in the file:

```
function sendTo(conn, message) {
  conn.send(JSON.stringify(message));
}
```

What this function does is ensure that all of our messages are always encoded in the JSON format. This also helps reduce the amount of code we have to write. It is always good practice to keep message sending in one place in case something else has to be done when sending messages to clients.

The last thing we have to do is provide a way to clean up client connections when they disconnect. Luckily, our library provides an event just when this happens. We can listen to this event and delete our user in this way:

```
connection.on(''close'', function () {
    if (connection.name) {
      delete users[connection.name];
    }
});
```

This should be added in the `connection` event as in the case of the message handler.

Now it is time to test our server with our `login` command. We can use the client as we did before to test out our `login` command. One thing to keep in mind is that messages we send now have to be encoded in the JSON format for them to be accepted by the server.

Once we connect, we can send the following message to our server:

```
{ ""type"": ""login"", ""name"": ""Foo"" }
```

The output you receive should look similar to this:

```
dan:webrtc-book/ (masterX) $ wscat -c ws://localhost:8888
connected (press CTRL+C to quit)
> { "type": "login", "name": "Foo" }
< {"type":"login","success":true}
>
```

Initiating a call

From here on, our code does not get any more complex than the `login` handler. We will create a set of handlers to pass our message correctly for each step of the way. One of the first calls that is made after logging in is the `offer` handler, which designates that one user would like to call another.

It is a good idea not to get call initiations mixed up with the `offer` step of WebRTC. In this example, we have combined the two to make our API easier to work with. In most settings, these steps will be separated. This can be seen in an application, such as Skype, where the other user has to accept the incoming call before a connection is established between the two users.

We can now add the `offer` handler into this code:

```
case ""offer"":
        console.log(""Sending offer to"", data.name);
        var conn = users[data.name];

        if (conn != null) {
          connection.otherName = data.name;
          sendTo(conn, {
            type: ""offer"",
            offer: data.offer,
            name: connection.name
          });
        }

        break;
```

The first thing we do is get `connection` of the user we are trying to call. This is easy to do since the ID of the other user is always where our `connection` is stored in our user-lookup object. We then check if the other user exists and if so, send them the details of `offer`. We also add an `otherName` property to the user's `connection` object so that we can look this up easily later on in the code. You might also notice that none of this code is WebRTC-specific. This could potentially refer to any sort of calling technology between two users. We will cover this in more detail later on in the chapter.

Something you may also notice is the lack of error handling. This is perhaps one of the most tedious parts of WebRTC. Since a call can fail at any point of the process, we have many places where making a connection can fail. It can also fail for various reasons, such as network availability, firewalls, and more. In this book, we leave it up to the user to handle each error case individually in the manner that they would like.

Answering a call

Answering the response is just as easy as `offer`. We follow a similar pattern and let the clients do most of the work. Our server will simply let any message pass through as `answer` to the other user. We can add this in after the `offer` handling case:

```
case ""answer"":
        console.log(""Sending answer to"", data.name);
        var conn = users[data.name];

        if (conn != null) {
          connection.otherName = data.name;
          sendTo(conn, {
            type: ""answer"",
            answer: data.answer
          });
        }

        break;
```

You can see how similar the code looks in the preceding listing. Note, we are also relying on `answer` to come from the other user. If a user were to send `answer` first, instead of `offer`, it could potentially mess up our server implementation. There are many use cases where this server will not be sufficient enough, but it will work well for integration during the next chapter.

This should be a good start to the `offer` and `answer` mechanism in WebRTC. You should see that it follows the `createOffer` and `createAnswer` functions on `RTCPeerConnection`. This is exactly where we will start plugging in our server connection to handle remote clients.

We can even test our current implementation using the WebSocket client we used before. Connecting two clients at the same time allows us to send `offer` and response between the two. This should give you more insight into how this will work in the end. You can see the results from running two clients simultaneously in the terminal window, as shown in the following screenshot:

In my case, my `offer` and `answer` were simple string messages. If you recall from *Chapter 3, Creating a Basic WebRTC Application*, in *The WebRTC API* section, we detailed the Session Description Protocol (SDP). This is what the `offer` and `answer` strings will actually be filled with when making a WebRTC call. If you do not remember what the SDP is, refer to the *Session Description Protocol* subsection under *The WebRTC API* section in *Chapter 3, Creating a Basic WebRTC Application*, and refresh your memory.

Handling ICE candidates

The final piece of the WebRTC signaling puzzle is handling ICE candidates between users. Here, we use the same technique as before to pass messages between users. The difference in the candidate message is that it might happen multiple times per user and in any order between the two users. Thankfully, our server is designed in a way that can handle this easily. You can add this `candidate` handler code to your file:

```
case ""candidate"":
    console.log(""Sending candidate to"", data.name);
    var conn = users[data.name];

    if (conn != null) {
      sendTo(conn, {
        type: ""candidate"",
```

```
                candidate: data.candidate
            });
        }

        break;
```

Since the call is already set up, we do not need to add the other user's name in this function either. Go ahead and test this one on your own using the terminal WebSocket client. It should work similarly to the `offer` and `answer` functions, passing messages between the two.

Hanging up a call

Our last bit is not part of the WebRTC specification, but is still a good feature to have—hanging up. This will allow our users to disconnect from another user so they are available to call someone else. This will also notify our server to disconnect any user references we have in our code. You can add the `""leave""` handler as detailed in the following code:

```
case ""leave"":
        console.log(""Disconnecting user from"", data.name);
        var conn = users[data.name];
        conn.otherName = null;

        if (conn != null) {
          sendTo(conn, {
            type: ""leave""
          });
        }

        break;
```

This will also notify the other user of the `leave` event so they can disconnect their peer connection accordingly. Another thing we have to do is handle the case of when a user drops their connection from the signaling server. This means we can no longer serve them and that we need to terminate their calls. We can change the `close` handler we used before to look similar to this:

```
connection.on(''close'', function () {
    if (connection.name) {
      delete users[connection.name];

      if (connection.otherName) {
```

```
        console.log(""Disconnecting user from"",
    connection.otherName);
        var conn = users[connection.otherName];
        conn.otherName = null;

        if (conn != null) {
          sendTo(conn, {
            type: ""leave""
          });
        }
      }
    }
  }
});
```

This will now disconnect our users if they happen to terminate their connection unexpectedly from the server. This can help in cases where we are still in `offer`, `answer`, or a `candidate` state but the other user closes their browser window. In this case, the WebRTC API will not send any events of this happening and we need another way to know that the user has left. Having the signaling server handle in this case helps make our application more reliable and stable overall.

Complete signaling server

Here is the entire code for our signaling server. This includes logging in and handling all response types. I also added a listening handler at the end to notify you when the server is ready to accept WebSocket connections:

```
var WebSocketServer = require(''ws'').Server,
    wss = new WebSocketServer({ port: 8888 }),
    users = {};

wss.on(''connection'', function (connection) {
  connection.on(''message'', function (message) {
    var data;

    try {
      data = JSON.parse(message);
    } catch (e) {
      console.log(""Error parsing JSON"");
      data = {};
    }
```

```
switch (data.type) {
  case ""login"":
    console.log(""User logged in as"", data.name);
    if (users[data.name]) {
      sendTo(connection, {
        type: ""login",
        success: false
      });
    } else {
      users[data.name] = connection;
      connection.name = data.name;
      sendTo(connection, {
        type: "login",
        success: true
      });
    }

    break;
  case "offer":
    console.log("Sending offer to", data.name);
    var conn = users[data.name];

    if (conn != null) {
      connection.otherName = data.name;
      sendTo(conn, {
        type: "offer",
        offer: data.offer,
        name: connection.name
      });
    }

    break;
  case "answer":
    console.log("Sending answer to", data.name);
    var conn = users[data.name];

    if (conn != null) {
      connection.otherName = data.name;
      sendTo(conn, {
        type: "answer",
        answer: data.answer
      });
    }
```

```
        break;
      case "candidate":
        console.log("Sending candidate to", data.name);
        var conn = users[data.name];

        if (conn != null) {
          sendTo(conn, {
            type: "candidate",
            candidate: data.candidate
          });
        }

        break;
      case "leave":
        console.log("Disconnecting user from", data.name);
        var conn = users[data.name];
        conn.otherName = null;

        if (conn != null) {
          sendTo(conn, {
            type: "leave"
          });
        }

        break;
      default:
        sendTo(connection, {
          type: "error",
          message: "Unrecognized command: " + data.type
        });

        break;
    }
  });

  connection.on('close', function () {
    if (connection.name) {
      delete users[connection.name];

      if (connection.otherName) {
        console.log("Disconnecting user from",
connection.otherName);
        var conn = users[connection.otherName];
        conn.otherName = null;
```

```
          if (conn != null) {
            sendTo(conn, {
              type: "leave"
            });
          }
        }
      }
    });
});

function sendTo(conn, message) {
  conn.send(JSON.stringify(message));
}

wss.on('listening', function () {
    console.log("Server started...");
});
```

 You may notice that we are using an unsecure WebSocket server in our example. In the real world, the WebSocket protocol actually supports SSL similar to how HTTP supports HTTPS. You can simply use wss:// to enable this when connecting the server.

Feel free to test our server application using the WebSocket client just as you did earlier. You can even try connecting three, four, or more users to the server and see how it handles multiple connections. You will probably also find many use cases that our server does not handle. It is a good idea to note these cases and even improve the server to work around them.

Signaling in the real world

It has taken us a lot of effort to get to a basic signaling server to connect two WebRTC users. At this point, you may be wondering how signaling servers are built in the real world for production applications. Since signaling is such an abstract concept that is not defined by the WebRTC specification, the answer is that anything goes.

Signaling is such a complex and difficult issue to solve because of the "anything goes" mentality it brings. There are many resources out there offered by the WebRTC makers, but none of them details how exactly signaling is best implemented for users. There are many issues to solve here and not all of them are the same for every use case. Some developers might need a highly scalable solution that can connect millions of users across the globe. Another developer might need a solution that integrates with Facebook, and another might need to integrate with Twitter. It is an extremely tough topic to cover and will require lots of time and research to find the best solution. Here, we will detail a few of the common pitfalls and solutions when researching signaling servers.

The woes of WebSockets

The great thing about WebSockets is that it has brought bidirectional communication to browsers. Many consider WebSockets to be the answer to all their problems, enabling faster socket connections directly to servers. This being said, there are still a few wrinkles to iron out in the WebSocket space.

One of these wrinkles is the **network firewall problem**. Under ideal conditions, WebSockets is a reliable connection but, unlike its HTTP counterpart, it is easy for it to become unstable under proxy configurations. The additional overhead of a **Virtual Private Network (VPN)** or complex firewall systems can cause the connection success rate to drop significantly. This means you will have to fall back on other technologies such as HTTP streaming to accomplish the same task.

This now introduces some race conditions in the WebRTC space. Any latency in the pipeline can cause out-of-order message processing, giving poor results when connecting in WebRTC. Remember that when making a WebRTC connection, the order is important and doing things out of order can cause a failed connection.

All this aside, WebSockets is an amazing piece of technology. The moral of the story is that when creating a real-world product there will be hiccups using WebSockets as a signaling server technology. Many companies are using them effectively today but have many fallbacks in cases where WebSockets does not work well in given network conditions.

Connecting with other services

One of the most exciting parts of WebRTC is that it will not only work well as a standalone solution, but also pairs well with other technologies. There have been numerous peer connectivity applications before WebRTC came around and since its introduction, efforts have been put forward to make WebRTC backward compatible. This includes using common frameworks seen in instant messaging systems and even technologies that our cellular phones use today.

XMPP

XMPP is an instant messaging protocol that dates back to the 90's under the name Jabber. The protocol was aimed at defining a common way to implement instant messaging, user presence, and contact lists. It is an open standard which anyone can use and integrate into their application. A large number of big instant messaging platforms have integrated XMPP into their service at some point, including Google Talk, Facebook Chat, and AOL Instant Messenger.

It is this wealth of historical data that makes XMPP an easy platform to use. It gives a lot of power to any typical WebRTC application since many video and audio communication platforms will at some point need presence and contact list data. On top of this, it is secure, has lots of documentation, and is extremely flexible in its implementation. There are a number of well-built ports to JavaScript and WebRTC as well as companies that are dedicated to offering WebRTC-based XMPP as a service. If you are able to get one working, it would be a great deal better than the simple signaling server we have built in this chapter.

Session Initiation Protocol

The **Session Initiation Protocol (SIP)** is another standard dating back to the 90's. It is a signaling protocol that has been designed targeting cellular networks and phone systems. It is a well-defined and extremely well-supported protocol seen in use by major cell networks and network equipment providers.

The aim of SIP integration and WebRTC is to provide communication support with SIP-based phone devices that do not have WebRTC support. If we made our connection to a server that could translate our information, it could easily connect to a mobile phone or other communication devices. If it used SIP, it would also come with support for many of the features that come with phones today.

SIP is another large topic in itself. You can find countless ports and resources on integrating SIP with WebRTC on the Web. As opposed to XMPP, this is definitely on the other end of the spectrum as far as difficulty and complexity is concerned. Phone-based communication is a completely different topic with its own technologies and standards. We will not cover this integration in the scope of this book, but feel free to stop and look for resources on your own.

Self-test questions

Q1. The goal of the signaling server is to connect two users on separate networks so that they can make a peer connection between them. True or False?

Q2. What technology does WebSockets use to make a bidirectional connection between client and server?

1. UDP
2. TCP
3. ICE
4. STUN

Q3. Using JSON for client-to-server messages gives us which of the following benefits?

1. String-based packet data for easy transport
2. Complex structure definition inside messages
3. Widely supported encoding and decoding methods
4. All of the above

Q4. In signaling, the order of operations is offer, answer, then sending candidates back and forth until a connection is made. True or False?

Summary

Over the course of this chapter, we covered each step of the signaling process. We walked through setting up a Node.js application, identifying users, and sending the entirety of offer/answer mechanisms between users. We also detailed disconnecting, leaving connections, and sending candidates between users.

You should now have a firm understanding of how a signaling server works. The server that we built is built to be simple and straightforward from a pure learning standpoint. We could fill an entire book up with new features that we could add to our server, such as authentications, buddy lists, and more. If you are feeling adventurous, feel free to add as many functionalities as you would like to our implementation.

We also covered a little about real-world signaling applications. This is just the tip of the iceberg as far as information about signaling is concerned. There are a wealth of use cases and implementations out there for WebRTC signaling. My recommendation would be to collect your requirements early and stick to the simplest solution that meets all of your requirements. This could be anything from a simple WebSocket server to the most complex SIP implementation.

In the next chapter, we will be integrating our server with an actual WebRTC client. This will enable us to make WebRTC connections between users at different locations. This is the beginning of a full-fledged WebRTC application that can be used to connect people all over the world.

Connecting Clients Together

5

Now that we have implemented our own signaling server, it is time to build an application to utilize its power. In this chapter, we are going to build a client application that allows two users on separate machines to connect and communicate in real time using WebRTC. By the end of this chapter, we will have a well-designed working example of what most WebRTC applications function like.

In this chapter, we will cover the following topics:

- Getting a connection to our server from a client
- Identifying users on each end of the connection
- Initiating a call between two remote users
- Hanging up a finished call

If you have not already completed *Chapter 4, Creating a Signaling Server*, now is a great time to go back and do so. This chapter is built upon the server which we built in that chapter, so you will have to know how to set up and run the server locally on your computer.

The client application

The goal of the client application is to enable two users to connect and communicate with each other from different locations. This is often seen as the *hello world* of WebRTC applications, and many examples of this type of application can be seen around the Web and at WebRTC-based conferences and events. Chances are you have used something much similar before to what we will build in this chapter.

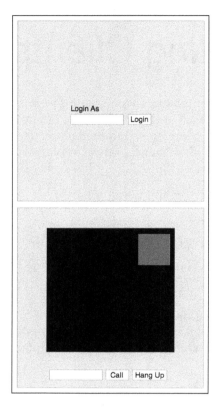

Our application will have two pages in it: one for selecting a username and the other for calling another user. Keep in mind that the page itself will be extremely simplistic in nature. We will mostly be focusing on how to build the actual WebRTC functionality. We will now look at the initial wireframe mock-ups to be used as a guideline before building the application.

You can say that it is not a complex application by any means. The two pages will be the `div` tags that we will switch out using JavaScript. Also, most input is done through simple event handlers. If you have done some basic HTML5 and JavaScript programming, the code in this chapter should look familiar.

What we are going to focus on is integrating our application with our signaling server. This will mean taking the events that we used locally in This will mean taking the events that we used locally in *The RTCPeerConnection object* subsection under *The WebRTC API* section in *Chapter 3, Creating a Basic WebRTC Application*, and sending them between two different pages, instead of using them on the same page. One way to test this is by opening two browser tabs so that both point to the same page and have them call each other.

Setting up the page

To start, we need to create a basic HTML page. The following is the boilerplate code used to give us something to work from. Copy this code into your own `index.html` document:

```html
<!DOCTYPE html>
<html lang="en">
  <head>
    <meta charset="utf-8" />

    <title>Learning WebRTC - Chapter 5: Connecting Clients
Together</title>

    <style>
      body {
        background-color: #3D6DF2;
        margin-top: 15px;
        font-family: sans-serif;
        color: white;
      }

      video {
        background: black;
        border: 1px solid gray;
      }

      .page {
        position: relative;
        display: block;
        margin: 0 auto;
        width: 500px;
```

```
        height: 500px;
      }

      #yours {
        width: 150px;
        height: 150px;
        position: absolute;
        top: 15px;
        right: 15px;
      }

      #theirs {
        width: 500px;
        height: 500px;
      }
    </style>
  </head>
  <body>
    <div id="login-page" class="page">
      <h2>Login As</h2>
      <input type="text" id="username" />
      <button id="login">Login</button>
    </div>

    <div id="call-page" class="page">
      <video id="yours" autoplay></video>
      <video id="theirs" autoplay></video>
      <input type="text" id="their-username" />
      <button id="call">Call</button>
      <button id="hang-up">Hang Up</button>
    </div>

    <script src="client.js"></script>
  </body>
</html>
```

Most of this should start looking familiar by now. We have identified two pages using the div tags that we will show and hide using the display property. On top of this, we have created several buttons and inputs for getting information from the user. Finally, you should recognize the two video elements used to display your video stream and the remote user's video stream. You can also add your own CSS to the page if you would like your example to look different from the default styles.

Do not forget to host this file using the static web server as shown earlier. Due to security restrictions present in most browsers, opening this file directly instead of navigating to a locally hosted server, will not allow you to make a WebRTC connection.

Getting a connection

The first thing we will do is establish a connection with our signaling server. The signaling server that we built in *Chapter 4, Creating a Signaling Server*, is entirely based on the WebSocket protocol. The great thing about the technology that we built on top of is that it requires no extra libraries to connect to the server. It simply uses the built-in power of WebSocket in most up-to-date browsers today. We can just create a WebSocket object directly, and connect to our server in no time at all.

The other wonderful thing about using WebSockets is that the standard is much further along in the process. On most browsers today, there are no checks or prefixes needed and it should be readily available. As always, however, check the latest documentation for your browser.

We can start by creating the client.js file that our HTML page includes. You can add the following connection code:

```
var name,
    connectedUser;

var connection = new WebSocket('ws://localhost:8888');

connection.onopen = function () {
  console.log("Connected");
};

// Handle all messages through this callback
connection.onmessage = function (message) {
  console.log("Got message", message.data);

  var data = JSON.parse(message.data);

  switch(data.type) {
    case "login":
      onLogin(data.success);
      break;
```

```
      case "offer":
        onOffer(data.offer, data.name);
        break;
      case "answer":
        onAnswer(data.answer);
        break;
      case "candidate":
        onCandidate(data.candidate);
        break;
      case "leave":
        onLeave();
        break;
      default:
        break;
    }
  };

  connection.onerror = function (err) {
    console.log("Got error", err);
  };

  // Alias for sending messages in JSON format
  function send(message) {
    if (connectedUser) {
      message.name = connectedUser;
    }

    connection.send(JSON.stringify(message));
  };
```

The initial thing we do is set up the connection to our server. We do this by passing in the location of our server, including the ws:// protocol prefix on our URI. Next, we set up a series of event handlers. The main one to take note of is the onmessage handler, which is where we will get all of our WebRTC-based messages. The switch method calls different functions based on the message type, which we will fill out later in this chapter. Finally, we create a simple send method, which will automatically attach the other user's ID to our messages and encode them for us. We also set up some variables to hold your username and the other user's ID for later use. When you open this file now, you should see a simple connection message:

The WebSocket API is a solid foundation for building real-time applications. As you will see from this chapter, it enables us to send back and forth information between the browser and the server almost instantly. Not only can we use this for signaling data, but also other information as well. WebSockets has been used in a number of different websites, such as multiplayer gaming, stock brokering, and more.

Logging in to the application

The first interaction with the server is to log in with a unique username. This is to identify ourselves as well as give other users a unique identifier to call us by. To do this, we simply send a name to the server, which will then tell us if the username has been taken or not. In our application, we will let the user select any name they would like.

To implement this, we need to add a bit of functionality to our application's script file. You can add the following to your JavaScript:

```javascript
var loginPage = document.querySelector('#login-page'),
    usernameInput = document.querySelector('#username'),
    loginButton = document.querySelector('#login'),
    callPage = document.querySelector('#call-page'),
    theirUsernameInput = document.querySelector('#their
username'),
    callButton = document.querySelector('#call'),
    hangUpButton = document.querySelector('#hang-up');

callPage.style.display = "none";

// Login when the user clicks the button
loginButton.addEventListener("click", function (event) {
  name = usernameInput.value;

  if (name.length > 0) {
    send({
      type: "login",
      name: name
    });
```

```
    }
  });

  function onLogin(success) {
    if (success === false) {
      alert("Login unsuccessful, please try a different name.");
    } else {
      loginPage.style.display = "none";
      callPage.style.display = "block";

      // Get the plumbing ready for a call
      startConnection();
    }
  };
```

Initially, we select some references to the elements on the page so that we can interact with the user and provide feedback in various ways. We then tell the `callPage` area to hide itself so that the user is just presented with the login flow. Then, we add a listener to the **Login** button so that when the user selects it, we send a message to the server telling it to log in. Finally, we implement the `onLogin` function that was referenced in our message callback earlier. If the login was successful, the application will show them the `callPage` area, and set up some of the requirements for making a WebRTC connection.

Starting a peer connection

The `startConnection` function is the first part of any WebRTC connection. Since this entire process is not reliant on another person to connect to, we can set this step up ahead, before the user has actually tried to call anyone. The steps included are:

1. Obtaining a video stream from the camera.
2. Verifying that the user's browser supports WebRTC.
3. Creating the `RTCPeerConnection` object.

This is implemented by the following JavaScript:

```
var yourVideo = document.querySelector('#yours'),
    theirVideo = document.querySelector('#theirs'),
    yourConnection, connectedUser, stream;

function startConnection() {
  if (hasUserMedia()) {
    navigator.getUserMedia({ video: true, audio: false }, function
(myStream) {
```

```
        stream = myStream;
        yourVideo.src = window.URL.createObjectURL(stream);

        if (hasRTCPeerConnection()) {
          setupPeerConnection(stream);
        } else {
          alert("Sorry, your browser does not support WebRTC.");
        }
      }, function (error) {
        console.log(error);
      });
    } else {
      alert("Sorry, your browser does not support WebRTC.");
    }
  }


function setupPeerConnection(stream) {
  var configuration = {
    "iceServers": [{ "url": "stun:stun.1.google.com:19302" }]
  };
  yourConnection = new RTCPeerConnection(configuration);

  // Setup stream listening
  yourConnection.addStream(stream);
  yourConnection.onaddstream = function (e) {
    theirVideo.src = window.URL.createObjectURL(e.stream);
  };

  // Setup ice handling
  yourConnection.onicecandidate = function (event) {
    if (event.candidate) {
      send({
        type: "candidate",
        candidate: event.candidate
      });
    }
  };
}


function hasUserMedia() {
  navigator.getUserMedia = navigator.getUserMedia ||
navigator.webkitGetUserMedia || navigator.mozGetUserMedia ||
navigator.msGetUserMedia;
  return !!navigator.getUserMedia;
```

```
}

function hasRTCPeerConnection() {
    window.RTCPeerConnection = window.RTCPeerConnection ||
window.webkitRTCPeerConnection || window.mozRTCPeerConnection;
    window.RTCSessionDescription = window.RTCSessionDescription ||
window.webkitRTCSessionDescription ||
window.mozRTCSessionDescription;
    window.RTCIceCandidate = window.RTCIceCandidate ||
window.webkitRTCIceCandidate || window.mozRTCIceCandidate;
    return !!window.RTCPeerConnection;
}
```

This should all look rather familiar by now. Much of this code has been copied from the examples explained in *Chapter 3, Creating a Basic WebRTC Application*. As always, check for the correct browser prefixes and handle any errors accordingly. If you run this code, your page should now allow you to log in and then seek permission to use the video camera stream on your device. Also, as you may remember, the audio is set to false to avoid loud audio feedback when testing the connection on the same device:

You should have something similar to the preceding screenshot after completing this section. If you have issues getting this far, go back and review the previous chapter to ensure your server is set up properly. Also, make sure you are hosting the file on a local server to ensure that the getUserMedia API is functioning properly.

Initiating a call

Now that we have set everything up properly, we are ready to initiate a call with a remote user. Sending the offer to another user starts all this. Once a user gets the offer, he/she will create a response and start trading ICE candidates, until he/she successfully connects. This process is exactly the same as mentioned in *Chapter 3, Creating a Basic WebRTC Application*, except for the fact that it is now being done remotely over our signaling server. To accomplish this, we add the following code to our script:

```
callButton.addEventListener("click", function () {
  var theirUsername = theirUsernameInput.value;

  if (theirUsername.length > 0) {
    startPeerConnection(theirUsername);
  }
});

function startPeerConnection(user) {
  connectedUser = user;

  // Begin the offer
  yourConnection.createOffer(function (offer) {
    send({
      type: "offer",
      offer: offer
    });
    yourConnection.setLocalDescription(offer);
  }, function (error) {
    alert("An error has occurred.");
  });
};

function onOffer(offer, name) {
  connectedUser = name;
  yourConnection.setRemoteDescription(new
RTCSessionDescription(offer));

  yourConnection.createAnswer(function (answer) {
    yourConnection.setLocalDescription(answer);
    send({
      type: "answer",
      answer: answer
    });
```

```
  }, function (error) {
    alert("An error has occurred");
  });
};

function onAnswer(answer) {
  yourConnection.setRemoteDescription(new
RTCSessionDescription(answer));
};

function onCandidate(candidate) {
  yourConnection.addIceCandidate(new RTCIceCandidate(candidate));
};
```

We start by adding a `click` handler to the **Call** button, which initiates the process. Then we implement a number of the functions expected by our message handler that is connected to our server. These will be processed asynchronously until both the parties have made a successful connection. Most of the work has already been done in the server and WebSocket layer, making the implementation of this part easier.

After running this code, you should be able to now log in to the server with two browser tabs with different usernames. You can then call the other tab using the call function which will successfully make a WebRTC connection between the clients.

Congratulations on making a fully-functioning WebRTC application! This is a major step to creating amazing peer-to-peer web-based applications. Something with this much power would typically take the span of several books and frameworks to get working, but we have done it in just a few short chapters with powerful technology.

Inspecting the traffic

Debugging a real-time application can be a tough process. With many things happening all at once, it is hard to build an entire picture of what is going on at any given moment. This is where using a modern browser with the WebSocket protocol really shines. Most browsers today will have some way to not only see the WebSocket connection to the server, but also inspect every packet that gets sent across the wire.

In my example, I use Chrome to inspect the traffic. Opening up the debugging tools by navigating to **View | Developer | Developer Tools**, will give me access to an array of tools for debugging web applications. Opening the **Network** tab will then show all of the network traffic that has been made by the page. If you do not see any network traffic, refresh the page with **Developer Tools** open. From there, the connection to **localhost** is easily seen on the list. When you select it, you have the option to see the frames that have been sent using the WebSocket connection. It shows every packet sent in human-readable format for easy debugging.

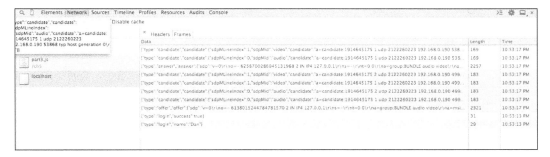

You should be able to see every step of the way in this view. The preceding screenshot shows the `login`, `offer`, `answer`, and every ICE candidate sent across the connection. This way, I can inspect each one for errors, such as malformed data, in the messages. When debugging web applications, it is a good idea to always utilize the built-in tools as much as possible.

There are also many other ways to get this information from the computer. Using the console output on both the server and client are great ways to get small pieces of information. You can also look into using network proxies and packet interceptors to intercept the packets being sent from the browser. This is much harder to set up, but will give much more information about the data being sent between the client and server. I will leave it up to the readers as an exercise to figure out the many ways of debugging web applications.

Hanging up a call

The last feature we will implement is the ability to hang up an in-progress call. This will notify the other user of our intention to close the call and stop transmitting information. It will take just a few additional lines to our JavaScript:

```
hangUpButton.addEventListener("click", function () {
  send({
    type: "leave"
  });

  onLeave();
});

function onLeave() {
  connectedUser = null;
  theirVideo.src = null;
  yourConnection.close();
  yourConnection.onicecandidate = null;
  yourConnection.onaddstream = null;
  setupPeerConnection(stream);
};
```

When the user clicks on the **Hang Up** button, it will send a message to the other user and destroy the connection locally. There are a few things required to successfully destroy the connection and also allow another call to be made in the future:

1. First off, we need to notify our server that we are no longer communicating.

2. Secondly, we need to tell RTCPeerConnection to close, and this will stop transmitting our stream data to the other user.

3. Finally, we tell the connection to set up again. This instantiates our connections to the open state so that we can accept new calls.

A complete WebRTC client

The following is the entire JavaScript code used in our client application. This includes all the code to hook up the UI, connect to the signaling server, and initiate a WebRTC connection with another user:

```
var connection = new WebSocket('ws://localhost:8888'),
    name = "";
```

```javascript
var loginPage = document.querySelector('#login-page'),
    usernameInput = document.querySelector('#username'),
    loginButton = document.querySelector('#login'),
    callPage = document.querySelector('#call-page'),
    theirUsernameInput = document.querySelector('#their-
username'),
    callButton = document.querySelector('#call'),
    hangUpButton = document.querySelector('#hang-up');

callPage.style.display = "none";

// Login when the user clicks the button
loginButton.addEventListener("click", function (event) {
  name = usernameInput.value;

  if (name.length > 0) {
    send({
      type: "login",
      name: name
    });
  }
});

connection.onopen = function () {
  console.log("Connected");
};

// Handle all messages through this callback
connection.onmessage = function (message) {
  console.log("Got message", message.data);

  var data = JSON.parse(message.data);

  switch(data.type) {
    case "login":
      onLogin(data.success);
      break;
    case "offer":
      onOffer(data.offer, data.name);
      break;
    case "answer":
```

```
        onAnswer(data.answer);
        break;
      case "candidate":
        onCandidate(data.candidate);
        break;
      case "leave":
        onLeave();
        break;
      default:
        break;
    }
  };

  connection.onerror = function (err) {
    console.log("Got error", err);
  };

  // Alias for sending messages in JSON format
  function send(message) {
    if (connectedUser) {
      message.name = connectedUser;
    }

    connection.send(JSON.stringify(message));
  };

  function onLogin(success) {
    if (success === false) {
      alert("Login unsuccessful, please try a different name.");
    } else {
      loginPage.style.display = "none";
      callPage.style.display = "block";

      // Get the plumbing ready for a call
      startConnection();
    }
  };

  callButton.addEventListener("click", function () {
    var theirUsername = theirUsernameInput.value;

    if (theirUsername.length > 0) {
      startPeerConnection(theirUsername);
    }
```

```
});

hangUpButton.addEventListener("click", function () {
  send({
    type: "leave"
  });

  onLeave();
});

function onOffer(offer, name) {
  connectedUser = name;
  yourConnection.setRemoteDescription(new
RTCSessionDescription(offer));

  yourConnection.createAnswer(function (answer) {
    yourConnection.setLocalDescription(answer);
    send({
      type: "answer",
      answer: answer
    });
  }, function (error) {
    alert("An error has occurred");
  });
}

function onAnswer(answer) {
  yourConnection.setRemoteDescription(new
RTCSessionDescription(answer));
}

function onCandidate(candidate) {
  yourConnection.addIceCandidate(new RTCIceCandidate(candidate));
}

function onLeave() {
  connectedUser = null;
  theirVideo.src = null;
  yourConnection.close();
  yourConnection.onicecandidate = null;
  yourConnection.onaddstream = null;
  setupPeerConnection(stream);
}
```

```
function hasUserMedia() {
  navigator.getUserMedia = navigator.getUserMedia ||
navigator.webkitGetUserMedia || navigator.mozGetUserMedia ||
navigator.msGetUserMedia;
  return !!navigator.getUserMedia;
}

function hasRTCPeerConnection() {
  window.RTCPeerConnection = window.RTCPeerConnection ||
window.webkitRTCPeerConnection || window.mozRTCPeerConnection;
  window.RTCSessionDescription = window.RTCSessionDescription ||
window.webkitRTCSessionDescription ||
window.mozRTCSessionDescription;
  window.RTCIceCandidate = window.RTCIceCandidate ||
window.webkitRTCIceCandidate || window.mozRTCIceCandidate;
  return !!window.RTCPeerConnection;
}

var yourVideo = document.querySelector('#yours'),
    theirVideo = document.querySelector('#theirs'),
    yourConnection, connectedUser, stream;

function startConnection() {
  if (hasUserMedia()) {
    navigator.getUserMedia({ video: true, audio: false }, function
(myStream) {
      stream = myStream;
      yourVideo.src = window.URL.createObjectURL(stream);

      if (hasRTCPeerConnection()) {
        setupPeerConnection(stream);
      } else {
        alert("Sorry, your browser does not support WebRTC.");
      }
    }, function (error) {
      console.log(error);
    });
  } else {
    alert("Sorry, your browser does not support WebRTC.");
  }
}

function setupPeerConnection(stream) {
  var configuration = {
```

```
      "iceServers": [{ "url": "stun:stun.1.google.com:19302" }]
    };
    yourConnection = new RTCPeerConnection(configuration);

    // Setup stream listening
    yourConnection.addStream(stream);
    yourConnection.onaddstream = function (e) {
      theirVideo.src = window.URL.createObjectURL(e.stream);
    };

    // Setup ice handling
    yourConnection.onicecandidate = function (event) {
      if (event.candidate) {
        send({
          type: "candidate",
          candidate: event.candidate
        });
      }
    };
  }

  function startPeerConnection(user) {
    connectedUser = user;

    // Begin the offer
    yourConnection.createOffer(function (offer) {
      send({
        type: "offer",
        offer: offer
      });
      yourConnection.setLocalDescription(offer);
    }, function (error) {
      alert("An error has occurred.");
    });
  };
```

If you are having issues getting the client running, be sure to look over this code a few times to make sure you have everything correctly copied. Another thing to look at is browser-specific implementations. There are minor nuances among browsers, so keep an eye on the console for any errors you may see.

Improving the application

What we have built over the course of this chapter is an adequate place to jump off into bigger and better things. It provides a baseline of features that almost every peer-to-peer communication application needs. From here, it is a matter of adding on common web application features to enhance the experience.

The login experience is one place to start improving the experience. There are a number of well-built services to enable user identification through common platforms like Facebook and Google. Integration with either of these APIs is easy and straightforward, and provides a great way to ensure that each user is unique. They also provide friend list capabilities so that the user has a list of people to call even if it is his/her first time using the application.

On top of this, the application will need to be foolproof to ensure the best possible experience. User input should be checked at each part of the way by both the client and server. Also, there are several places where the WebRTC connection can fail, such as not supporting the technology, not being able to traverse firewalls, and not having enough bandwidth to stream a video call. A wealth of work has gone into making common phone communication platforms stable to avoid dropped calls, and the same amount of work will have to go into making any WebRTC platform stable.

Self-test questions

Q1. Creating a WebSocket connection in most browsers requires the installation of several frameworks to get working properly. True or false?

Q2. Which technology or technologies does the user's browser need to support to successfully run the example created in this chapter?

1. WebRTC
2. WebSockets
3. Media Capture and Streams
4. All of the above

Q3. The application allows for more than two users to connect with each other in a video call. True or false?

Q4. The best way to make our application more stable, resulting in fewer errors when trying to establish a call, is to add:

1. More CSS styles
2. Facebook login integration
3. More error checking and validation along every step of the way
4. Really cool looking animations

Summary

After completing this chapter, you should take a step back and congratulate yourself for making it this far. Over the course of this chapter, we have brought the entire first half of the book into perspective with a full-fledged WebRTC application. With how complex peer-to-peer connections can be, it is amazing that we have been able to make one successfully in just five short chapters. You can now put down that chat client and use your own hand-built solution for communicating with people all over the world!

Now you should have a grasp on the overall architecture of any WebRTC application. We have covered the implementation of not only the client, but also the supporting signaling server. We have even integrated other HTML5 technologies, such as WebSockets, to help us in making a remote peer-to-peer connection.

If there was any point where you can take a break and put down this book, now would be the time. This application is a jumping off point to start prototyping your own WebRTC application and adding new innovative features. After reading this far, it is also a good idea to research other WebRTC applications on the Web and the approach they took when developing. With a broad understanding of the inner workings of WebRTC applications, you should be able to learn a lot from looking at other open source examples on the Web.

In the upcoming chapters, we are going to expand and touch on many advanced topics when creating peer-to-peer applications on the Web. Audio and video calls are just the tipping point of what you can do with WebRTC ; we will explore the many other features of this technology. We will also cover how to make a more robust application by connecting with multiple users, the concerns when moving to mobile phones, and security of WebRTC applications.

Sending Data with WebRTC

6

Up to this point, we have focused solely on the audio and video capabilities of WebRTC. However, there is an entire subject that we have not even started to talk about — **arbitrary data**. It turns out that WebRTC is good at transferring data; not just audio and video streams, but any data we might have.

In this chapter, we are going to cover the WebRTC Data Channel Protocol and how it can be utilized in our communication application. Over the course of this chapter, we will cover the following topics:

- How the data channel fits into the WebRTC puzzle
- How to create a data channel object from a peer connection
- What are the encryption and security concerns
- What are the potential use cases for the data channel

Stream Control Transmission Protocol and data transportation

As it turns out, sending data over a peer connection is somewhat of a hard task. Typically, sending data between users today is done through the use of a strict TCP connection. This means using technologies such as AJAX and WebSockets to send data to a server and then receive it at the other end. This can often be a slow and cumbersome issue for high-performance applications. To transfer data between the two users, the developer has to pay for a server network to do this for them. This can be expensive since even a small distributed network of servers can cost thousands of dollars a month.

In our WebRTC model, we have already covered the need for a higher speed, lower latency connection between users. With this connection, we have the ability to send audio and video data quickly between our peers. The protocol used today for audio and video data, however, is specifically designed for frames of a video and audio stream. This is why WebRTC introduces the **Stream Control Transmission Protocol (SCTP)** as a way to send data on top of our currently setup peer connection.

The SCTP is yet another technology that sits on the independent WebRTC stack. This is another reason why WebRTC is such an interesting technology. It has brought many new methods of data transportation to every JavaScript developers' fingertips. **SCTP** sits on top of the **Datagram Transport Layer Security (DTLS)** protocol that is implemented for every WebRTC connection and provides an outlet for the **data channel** to bind to for data transportation. All of these sit on top of the **UDP** stack that is used as the base transportation method for all the WebRTC data that we discussed in *Chapter 3, Creating a Basic WebRTC Application*. So far, our stack currently looks something like this:

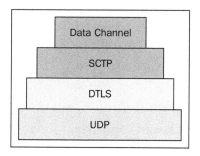

This may look like a lot of complexity but it is all for the benefit of the power of SCTP. The designers of WebRTC realized that every application would be unique in how it wants to harness the power of the data channel. Some might want the reliable delivery of TCP, while others might need the high performance of UDP. It was this issue that caused them to choose SCTP, which gives the best of both the worlds. The following are the features of SCTP:

- The transport layer is secured because it is built on top of DTLS
- The transport layer can run in reliable or unreliable modes
- The transport layer can guarantee or not guarantee the order of data packets
- The transmission of data is message oriented, allowing a message to be broken down and reassembled on the other side
- The transport layer provides both flow and congestion control

Some of these bullet points should sound really familiar to both UDP and TCP. This is because SCTP was born out of the limitations of both the protocols. It was designed as a way to fix the issues of TCP, but harness the power of UDP-like transports!

The SCTP specification is defined by using multiple **endpoints**, which send **messages** broken down through **chunks**. Here is a list of what these are:

- Endpoints: These are defined as any number of connections between two IP locations
- Messages: These are any portions of data sent to the SCTP layer from the application
- Chunks: These are packets of data ready to be sent across the wire and can represent a part of any message

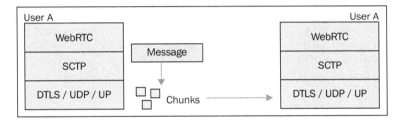

The entirety of the protocol spans 16 sections in all, but this should be a good enough start to understand how WebRTC works. The biggest thing to remember from this section is that the data channel takes a completely different route than the other data-based transport layers in the browser. It is also configurable and efficient in how it handles this data transportation. This is one of the many powerful technologies that WebRTC opens up for every JavaScript application developer.

The RTCDataChannel object

Now that we understand the underlying technologies at play, we can start learning how the actual RTCDataChannel object API works. The great thing is that this API is much less complex than the underlying workings of the SCTP. The main function to create a channel comes from an already established RTCPeerConnection object:

```
var peerConnection = new RTCPeerConnection();

// Establish your peer connection using signaling here

var dataChannel = peerConnection.createDataChannel("myLabel",
dataChannelOptions);
```

This is all you need to get started! The WebRTC API will take care of everything else on the browser's internal layer. This will all happen once signaling has been performed and the connection has been established. You can create a data channel at any point in the process until the RTCPeerConnection object is closed. Though it is closed, it will throw an error when trying to create a new channel.

There are a number of states that a data channel can be in:

- **connecting**: This is the default state, where a data channel waits for a connection
- **open**: This is the state when a connection is established and communication is possible
- **closing**: This is the state when a channel is currently being deconstructed
- **closed**: This is the state when a channel is closed and communication is not possible

The way that the browser notifies the application about the data channel states is through a series of events. When the other peer creates a channel, the ondatachannel event is fired on the RTCPeerConnection object, notifying that a channel has been created. Then the RTCDataChannel object itself has a few events to notify when it is opened, closed, has an error, or gets a message:

```
dataChannel.onerror = function (error) {
  console.log("Data channel error:", error);
};

dataChannel.onmessage = function (event) {
  console.log("Data channel message:", event.data);
};

dataChannel.onopen = function () {
  console.log("Data channel opened, ready to send messages!");
  dataChannel.send("Hello World!");
};

dataChannel.onclose = function () {
  console.log("Data channel has been closed.");
};
```

These events are straightforward and do exactly what they sound like. When creating a data channel, you should always wait for the onopen event to fire before sending any messages. Sending messages before the channel is open causes it to throw an error, stating that the data channel is not yet ready to send messages.

If these look anything similar to the WebSocket protocol that we used earlier, it is because it was intentionally made to look similar. Since the standards for WebSocket are well defined and understood, the same pattern was followed with the data channel to make it easy for developers to use. Do not be fooled, however, as WebSockets take an entirely different path in the browser for sending data to a remote location.

Data channel options

You may have noticed a `dataChannelOptions` object passed in our example. Since SCTP gives many different configurable ways to send data to the other peer, these options have been made available to the end developer. The options passed is optional and should be a regular JavaScript object:

```
var dataChannelOptions = {
  reliable: false,
  maxRetransmitTime: 3000
};
```

These options should vary between the benefits of UDP and TCP. Some settings will make the channel more reliable, while others will make it perform faster. These options include:

- `reliable`: This option shows whether message delivery is guaranteed or not.

- `ordered`: This shows whether messages should be received in the order they are sent.

- `maxRetransmitTime`: This option shows the maximum time to resend a failed message delivery.

- `maxRetransmits`: This option shows the maximum number of times to try to resend a failed message.

- `protocol`: This option will allow the forcing of a different subprotocol but will show error if the protocol is not supported by the user agent.

- `negotiated`: This option shows whether the developer is taking responsibility of creating a data channel on both peers or the browser should perform this step automatically. We will cover this more later on.

- `id`: This option is an identification string for the channel, which helps in negotiating multiple channels.

Although it seems like a lot, most of these options will be for advanced application usage. The main options are `reliable` and `ordered`, which will make the data channel act more like TCP if they are `true` and more like UDP if they are `false`.

The `negotiated` parameter is a setting for solving the synchronization problem in creating data channels for both users. This has to deal with the `ondatachannel` event that is fired on the `RTCPeerConnection` object. This defaults to `false`, meaning the browser will automatically fire an event on the other peer's side, notifying him/her of a new data channel. If set to `true`, the developer is taking responsibility of creating a data channel with the same ID on both the peers.

Sending data

The `send` method of the data channel is overloaded, much like the one on WebSockets. This allows many different JavaScript types to be sent over the transport layer. Using a different data type can help speed up the performance of an application. This is due to the heaviness of most string-based encodings, which require more chunks to send data. Currently, data channel supports the following types:

- `String`: A basic JavaScript string
- `Blob`: A data type introduced to represent a file-like format of raw data
- `ArrayBuffer`: A typed-array of fixed length that cannot be modified
- `ArrayBufferView`: A view frame of an underlying `ArrayBuffer`

Simply drop the variable in the `send` function and the browser will figure out the rest. You can then identify the type on the other side by testing its type:

```
dataChannel.onmessage = function (event) {
  var data = event.data;

  if (data instanceof Blob) {
    // handle Blob
  } else if (data instanceof ArrayBuffer) {
    // handle ArrayBuffer
  } else if (data instanceof ArrayBufferView) {
    // handle ArrayBufferView
  } else {
    // handle string
  }
};
```

Encryption and security

Having messages transmitted in a secure way is one of high importance for the designers of the WebRTC protocol. The reasoning behind this is that many large companies would not consider using a WebRTC application without the right level of security implementation. To get the widest adoption rate possible meant that security had to be implemented right into the design of the API itself.

One thing that you will notice in working with WebRTC is that encryption has been made a mandatory process for all implementations of the protocol. This means that each and every peer connection created between browsers is automatically using a good level of security. The encryption technology used had to satisfy several requirements for being used in peer applications:

- Messages should not be readable if they are stolen while in transit between peers

- A third party should not be able to forge messages to look as if they were sent by the connected peer

- Messages should not be editing-enabled while in transit to the other peer

- The encryption algorithm should be fast to support the highest bandwidth possible between the clients

The technology that was chosen to satisfy these requirements was the **DTLS**. You may recognize the last three words, often known as TLS to web developers. The TLS protocol dates back to its beginning in 1999 and is one of the most widely used security protocols on the Web today. It gained its popularity by being able to be inserted between the application and the transport layer, meaning you could secure an application with little or no changes to the logic of the application. The only drawback of using TLS is that it must be built on top of a TCP-based application, which the WebRTC protocol is not working on top of.

This is where DTLS comes in. It stems from the desire to have a simple and easy-to-use protocol like TLS, but with the power of the UDP transport layer. It takes after the TLS protocol using as many of the same concepts as possible while adding in support for UDP.

The biggest difference between TCP and UDP is that the former is able to guarantee that a message has made it to the other side. This happens to also be the major difference between TLS and DTLS. DTLS takes account for the ability to lose messages and receive them in the wrong order, making it possible to use DTLS with UDP. This makes DTLS a powerful choice for encrypting applications with requirements such as WebRTC.

The biggest takeaway here is to know that data is secure over the data channel and in WebRTC applications, in general. Steps have been taken to ensure that data cannot be modified in the middle of a peer-to-peer connection. This should work for most general peer-to-peer applications. The DTLS specification can be found at `https:// tools.ietf.org/html/rfc4347` to ensure that the security provided by DTLS is the right fit for your application.

Adding text-based chat

Now we will use what we have learned in this chapter to add data channel support to our communication application. Since the data channel can be used for any arbitrary data, we are going to add text-based chat as another feature of our application. The users will have a text box to enter a message into and a display of all the messages in the current call. When we are finished, it will end up looking something similar to this:

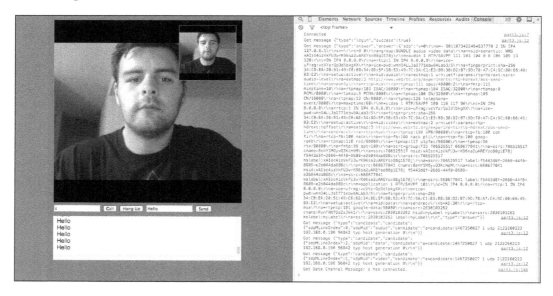

To get started, we will change the `call` page of our application. We will add three new elements—an input text field, a button, and a div. The input area and button will allow the user to enter text while `div` will hold all the messages between each user.

```
<div id="call-page" class="page">
    <video id="yours" autoplay></video>
    <video id="theirs" autoplay></video>
    <input type="text" id="their-username" />
    <button id="call">Call</button>
    <button id="hang-up">Hang Up</button>

    <input type="text" id="message"></input>
    <button id="send">Send</button>
    <div id="received"></div>
</div>
```

Next, we will add some JavaScript to our page to add the data channel functionality:

```
function openDataChannel() {
  var dataChannelOptions = {
    reliable: true
  };
  dataChannel = yourConnection.createDataChannel("myLabel",
dataChannelOptions);

  dataChannel.onerror = function (error) {
    console.log("Data Channel Error:", error);
  };

  dataChannel.onmessage = function (event) {
    console.log("Got Data Channel Message:", event.data);

    received.innerHTML += "recv: " + event.data + "<br />";
    received.scrollTop = received.scrollHeight;
  };

  dataChannel.onopen = function () {
    dataChannel.send(name + " has connected.");
  };

  dataChannel.onclose = function () {
    console.log("The Data Channel is Closed");
  };
}
```

This script sets up the data channel based on our connection. We will need to call this after the yourConnection variable has been instantiated and is ready to use. You can add the call to this function right after the yourConnection variable has been created. This code will setup a series of listeners for our data channel:

- onerror: This listener will catch any connection issues detected
- onmessage: This listener will receive messages from the other user
- onopen: This listener will tell us when the other user has connected
- onclose: This listener will tell us when the other user disconnects

At this point in time, we are just notifying the developer in the console. It is an exercise for the reader to make these listeners more useful!

 At the time of writing this book, I had to add {optional: [{RtpDataChannels: true}] to the RTCDataChannel call to get this example working in Chrome. The support for data channels is still in progress and might require some additional setup. If things do not seem to be working, search for data channels and your choice of browser to see whether any additional setup is required.

Now, we can add in an event listener when the user clicks on the **Send** button:

```
// Bind our text input and received area
sendButton.addEventListener("click", function (event) {
  var val = messageInput.value;
  received.innerHTML += "send: " + val + "<br />";
  received.scrollTop = received.scrollHeight;
  dataChannel.send(val);

});
```

This listener will grab the value of the text field. It will then add it to our message box locally. You might remember that we can run our data channel in a reliable state to ensure that both the users' message boxes stay in sync. This is the reason why we do not have to check whether the other user has received our message, since our data channel will take care of that for us. Finally, we tell the data channel to send the message which will fire the other user's onmessage handler, giving them the message data.

The message area also has a bit of style added to it to give it the scrolling message window look and feel, found in most communication applications:

```
#received {
    display: block;
    width: 480px;
    height: 100px;
    background: white;
    padding: 10px;
    margin-top: 10px;
    color: black;
    overflow: scroll;
}
```

Now, run the preceding code and you should have a working text-based chat application! Both users will be able to see, hear, and chat back and forth with each other, giving it the features of most popular communication applications. Unfortunately, we did not add any validation to the text, making our application easy to exploit in various ways. To make our application more robust, you may want to take a moment to improve our implementation with various features of your choice.

Use cases

This is just the tip of the iceberg when using the WebRTC data channel. Adding text-based chat is the easiest extension of the protocol with what we have learned so far. Since the specification is just the arbitrary data between two users, the possibilities are almost limitless with what you can do.

Gaming is one of the first logical extensions of the data channel protocol. In fact, peer-to-peer networking has been a common sight in many multiplayer games. Previously, games such as Quake 3 relied on peer-to-peer networks to send data between players when playing a game. The reasons behind this were the server cost and speed of transport.

With each player paying the expense of their Internet connection, this meant that the company did not have to implement complex servers to transfer data back and forth. Games could be played on any network where computers could reach each other over the Internet. It also gave the fastest speed of transportation at a time when local area networks outperformed the slow dial-up connections of the time. This trend has been increasingly changing over time, however, with the need to eliminate users from modifying the game's complex data and getting ahead of other players.

Many games today do use the data transfer capabilities of peer-to-peer networks to transfer game updates. The ability to transfer files in the browser comes from the concepts put forth by torrent-based file sharing networks. The concept behind sharing files over peer-to-peer networks is that a user's Internet connection can be used to give that file to other users in the network, much like a server would. This means that the transfer rate is only limited by the Internet connection of the user and how many users own the file, instead of the availability and cost of some central server. This is becoming an increasingly popular way to transfer large amounts of data among many users. The following diagram is an illustration of a common network layout for peer-to-peer games:

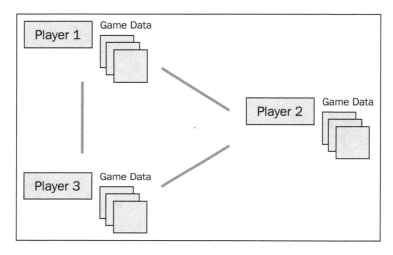

This concept is even being extended to the content delivery of files in the browser. Some developers are innovating on the idea of delivering scripts, images, and more in the browser by using the power of WebRTC. The idea here is that you can transfer these files between users using WebRTC connections instead of paying the cost of a large-scale **Content Delivery Network (CDN)**.

CDNs work by placing hundreds of servers across the world. This means that the chance of any user being close to a server on the network is high, giving high rates of file transfer. When you upload a file to the network, it is replicated between all the nodes across the world making it available at a moment's notice. The downside to this, however, is the high cost of maintaining this network.

This is where WebRTC can come in and help alleviate the situation. If the user could download data from a nearby user instead of a server, the network acts in a much similar manner. Once the file is downloaded, the user checks the file against a server using an algorithm to ensure that the contents match with what they should be expecting. This keeps the integrity of the files intact while making the network potentially larger than a CDN at a fraction of the cost.

It is these kinds of ideas that really provide the power behind the WebRTC Data Channel. This new form of data transfer opens the door to many network features that were not available to developers of web pages. There are many more innovative ideas to discover in every industry that can utilize the power of data channels.

Self-test questions

Q1. The data channel is not encrypted or secured at all, making it easy for hackers to modify the data being sent between users. True or false?

Q2. Which is not a correct state that `RTCDataChannel` can be in?

1. reconnecting
2. closed
3. connecting
4. open

Q3. The data channel can be run in reliable, unreliable, ordered, and unordered modes, giving it robust data transfer capabilities. True or false?

Q4. The biggest reason why TLS is not used in a WebRTC application is because of how hard it is to implement. True or false?

Q5. Common use cases for `RTCDataChannel` could include:

1. Multiplayer gaming
2. File transferring
3. Delivering content
4. All of the above

Summary

Upon finishing this chapter, you should have a firm grasp of yet another piece of the WebRTC puzzle—data transfer. It is an often overlooked, yet extremely powerful part of the WebRTC specification that can bring new light to web-based applications. In this chapter, we have used the data channel to bring text-based chat to our currently running WebRTC demo.

Now, you can officially say goodbye to our communication application, as this is the last time you will see it while reading this book. It is quite an amazing feat that we built over the course of just a few hundred pages. Not only have we built a multiuser application, but also one with many of the features found in popular communication applications built and used all over the world. This is really the power of not only web applications but also WebRTC.

If you have not done so already, this is yet another great opportunity to put down this book and start experimenting with our example. From here, there are a number of features missing from our example that can be added with a few hours of implementation. The first easy jump would be adding validation around the text input for each user. From there, it is easy to add the transfer of any data, even building a game that each player can control.

In the upcoming chapters, we are going to start diving into more advanced applications and examples of WebRTC. This will cover topics such as actually sharing files between two users, developing WebRTC application for mobiles and also building networks with more than two users.

7
File Sharing

The real power of a data channel comes when combining it with other powerful technologies from a browser. By opening up the power to send data peer-to-peer and combining it with a File API, we could open up all new possibilities in your browser. This means you could add file sharing functionalities that are available to any user with an Internet connection. Throughout the course of this chapter, we will focus on creating a simple file sharing application using the power of both the File API and the WebRTC Data Channel.

The application that we will build will be a simple one with the ability to share files between two peers. The basics of our application will be real-time, meaning that the two users have to be on the page at the same time to share a file. There will be a finite number of steps that both users will go through to transfer an entire file between them:

1. User A will open the page and type a unique ID.
2. User B will open the same page and type the same unique ID.
3. The two users can then connect to each other using `RTCPeerConnection`.
4. Once the connection is established, one user can select a file to share.
5. The other user will be notified of the file that is being shared, where it will be transferred to their computer over the connection and they will download the file.

The main thing we will focus on throughout the chapter is how to work with the data channel in new and exciting ways. We will be able to take the file data from the browser, break it down into pieces, and send it to the other user using only the `RTCPeerConnection` API. The interactivity that the API promotes will stand out in this chapter and can be used in a simple project that can easily be expanded after reading through the book.

Getting a file with the File API

One of the first things that we will cover is how to use the File API to get a file from the user's computer. There is a good chance you have interacted with the File API on a web page and have not even realized it yet! The API is usually denoted by the Browse or **Choose File** text located on an input field in the HTML page and often looks something similar to this:

Although the API has been around for quite a while, the one you are probably familiar with is the original specification, dating back as far as 1995. This was the *Form-based File Upload in HTML* specification that focused on allowing a user to upload a file to a server using an HTML form. Before the days of the file input, application developers had to rely on third-party tools to request files of data from the user. This specification was proposed in order to make a standard way to upload files for a server to download, save, and interact with. The original standard focused entirely on interacting with a file via an HTML form, however, and did not detail any way to interact with a file via JavaScript. This was the origin of the File API.

Fast-forward to the groundbreaking days of HTML5 and we now have a fully-fledged File API. The goal of the new specification was to open the doors to file manipulation for web applications, allowing them to interact with files similar to how a native-installed application would. This means providing access to not only a way for the user to upload a file, but also ways to read the file in different formats, manipulate the data of the file, and then ultimately do something with this data.

Although there are many great features of the API, we are going to only focus on one small aspect of this API. This is the ability to get binary file data from the user by asking them to upload a file. A typical application that works with files, such as Notepad on Windows, will work with file data in pretty much the same way. It asks the user to open a file in which it will read the binary data from the file and display the characters on the screen. The File API gives us access to the same binary data that any other application would use in the browser.

This is the great thing about working with the File API: it works in most browsers from a HTML page; similar to the ones we have been building for our WebRTC demos. To start building our application, we will put together another simple web page. This will look similar to the last ones, and should be hosted with a static file server as done in the previous examples. By the end of the book, you will be a professional single page application builder! Now let's take a look at the following HTML code that demonstrates file sharing:

```
<!DOCTYPE html>
<html lang="en">
  <head>
    <meta charset="utf-8" />

    <title>Learning WebRTC - Chapter 7: File Sharing</title>

    <style>
      body {
        background-color: #404040;
        margin-top: 15px;
        font-family: sans-serif;
        color: white;
      }

      .thumb {
        height: 75px;
        border: 1px solid #000;
        margin: 10px 5px 0 0;
      }

      .page {
        position: relative;
        display: block;
        margin: 0 auto;
        width: 500px;
        height: 500px;
      }
```

```
      #byte_content {
        margin: 5px 0;
        max-height: 100px;
        overflow-y: auto;
        overflow-x: hidden;
      }

      #byte_range {
        margin-top: 5px;
      }
    </style>
  </head>
  <body>
    <div id="login-page" class="page">
      <h2>Login As</h2>
      <input type="text" id="username" />
      <button id="login">Login</button>
    </div>

    <div id="share-page" class="page">
      <h2>File Sharing</h2>

      <input type="text" id="their-username" />
      <button id="connect">Connect</button>
      <div id="ready">Ready!</div>

      <br />
      <br />

      <input type="file" id="files" name="file" /> Read bytes:
      <button id="send">Send</button>
    </div>

    <script src="client.js"></script>
  </body>
</html>
```

The page should be fairly recognizable at this point. We will use the same page showing and hiding via CSS as done earlier. One of the main differences is the appearance of the file input, which we will utilize to have the user upload a file to the page. I even picked a different background color this time to spice things up.

Setting up our page

To start our code, we will build upon our application from the previous chapter. You can copy the files into a new folder for our file sharing application and add the HTML shown in the preceding section. You will also need all the steps from our JavaScript file to log in two users, create a WebRTC peer connection, and create a data channel between them. Copy the following code into your JavaScript file to get the page set up:

```javascript
var name,
connectedUser;

var connection = new WebSocket('ws://localhost:8888');

connection.onopen = function () {
  console.log("Connected");
};

// Handle all messages through this callback
connection.onmessage = function (message) {
  console.log("Got message", message.data);

  var data = JSON.parse(message.data);

  switch(data.type) {
    case "login":
      onLogin(data.success);
      break;
    case "offer":
      onOffer(data.offer, data.name);
      break;
    case "answer":
      onAnswer(data.answer);
      break;
    case "candidate":
      onCandidate(data.candidate);
      break;
    case "leave":
      onLeave();
      break;
    default:
```

```
      break;
  }
};

connection.onerror = function (err) {
  console.log("Got error", err);
};

// Alias for sending messages in JSON format
function send(message) {
  if (connectedUser) {
    message.name = connectedUser;
  }

  connection.send(JSON.stringify(message));
};

var loginPage = document.querySelector('#login-page'),
  usernameInput = document.querySelector('#username'),
  loginButton = document.querySelector('#login'),
  theirUsernameInput = document.querySelector('#their-
username'),
  connectButton = document.querySelector('#connect'),
  sharePage = document.querySelector('#share-page'),
  sendButton = document.querySelector('#send'),
  readyText = document.querySelector('#ready'),
  statusText = document.querySelector('#status');

sharePage.style.display = "none";
readyText.style.display = "none";

// Login when the user clicks the button
loginButton.addEventListener("click", function (event) {
  name = usernameInput.value;

  if (name.length > 0) {
    send({
      type: "login",
      name: name
    });
  }
});
```

```
function onLogin(success) {
  if (success === false) {
    alert("Login unsuccessful, please try a different name.");
  } else {
    loginPage.style.display = "none";
    sharePage.style.display = "block";

    // Get the plumbing ready for a call
    startConnection();
  }
};

var yourConnection, connectedUser, dataChannel, currentFile,
currentFileSize, currentFileMeta;

function startConnection() {
  if (hasRTCPeerConnection()) {
    setupPeerConnection();
  } else {
    alert("Sorry, your browser does not support WebRTC.");
  }
}

function setupPeerConnection() {
  var configuration = {
    "iceServers": [{ "url": "stun:stun.1.google.com:19302 " }]
  };
  yourConnection = new RTCPeerConnection(configuration, {optional:
[]});

  // Setup ice handling
  yourConnection.onicecandidate = function (event) {
    if (event.candidate) {
      send({
        type: "candidate",
        candidate: event.candidate
      });
    }
  };

  openDataChannel();
}
```

```javascript
function openDataChannel() {
  var dataChannelOptions = {
    ordered: true,
    reliable: true,
    negotiated: true,
    id: "myChannel"
  };
  dataChannel = yourConnection.createDataChannel("myLabel",
dataChannelOptions);

  dataChannel.onerror = function (error) {
    console.log("Data Channel Error:", error);
  };

  dataChannel.onmessage = function (event) {
    // File receive code will go here
  };

  dataChannel.onopen = function () {
    readyText.style.display = "inline-block";
  };

  dataChannel.onclose = function () {
    readyText.style.display = "none";
  };
}

function hasUserMedia() {
  navigator.getUserMedia = navigator.getUserMedia ||
navigator.webkitGetUserMedia || navigator.mozGetUserMedia ||
navigator.msGetUserMedia;
  return !!navigator.getUserMedia;
}

function hasRTCPeerConnection() {
  window.RTCPeerConnection = window.RTCPeerConnection ||
window.webkitRTCPeerConnection || window.mozRTCPeerConnection;
  window.RTCSessionDescription = window.RTCSessionDescription ||
window.webkitRTCSessionDescription ||
window.mozRTCSessionDescription;
  window.RTCIceCandidate = window.RTCIceCandidate ||
window.webkitRTCIceCandidate || window.mozRTCIceCandidate;
  return !!window.RTCPeerConnection;
}
```

```
function hasFileApi() {
  return window.File && window.FileReader && window.FileList &&
window.Blob;
}

connectButton.addEventListener("click", function () {
  var theirUsername = theirUsernameInput.value;

  if (theirUsername.length > 0) {
    startPeerConnection(theirUsername);
  }
});

function startPeerConnection(user) {
  connectedUser = user;

  // Begin the offer
  yourConnection.createOffer(function (offer) {
    send({
      type: "offer",
      offer: offer
    });
    yourConnection.setLocalDescription(offer);
  }, function (error) {
    alert("An error has occurred.");
  });
};

function onOffer(offer, name) {
  connectedUser = name;
  yourConnection.setRemoteDescription(new
RTCSessionDescription(offer));

  yourConnection.createAnswer(function (answer) {
    yourConnection.setLocalDescription(answer);

    send({
      type: "answer",
      answer: answer
    });
  }, function (error) {
    alert("An error has occurred");
  });
```

```
};

function onAnswer(answer) {
  yourConnection.setRemoteDescription(new
RTCSessionDescription(answer));
};

function onCandidate(candidate) {
  yourConnection.addIceCandidate(new RTCIceCandidate(candidate));
};

function onLeave() {
  connectedUser = null;
  yourConnection.close();
  yourConnection.onicecandidate = null;
  setupPeerConnection();
};
```

A lot of this should feel similar to the code you have seen throughout the rest of the book. We set up references to our elements on the screen as well as get the peer connection ready to be processed. When the user decides to log in, we send a login message to the server that we built in *Identifying users* section in *Chapter 4, Creating a Signaling Server*. The server will return with a success message telling the user they are logged in.

From here, we allow the user to connect to another WebRTC user who is given their username. This sends offer and response, connecting the two users together through the peer connection. Once the peer connection is created, we connect the users through a data channel so that we can send arbitrary data across.

Hopefully, this is pretty straightforward and you are able to get this code up and running in no time. It should all be familiar to you and if you have any questions, you can refer to the respective chapters on what you are missing. This is the last time we are going to refer to this code, so get comfortable with it before moving on!

Getting a reference to a file

Now that we have a simple page up and running, we can start working on the file sharing part of the application. The first thing the user needs to do is select a file from their computer's filesystem. This is easily taken care of already by the input element on the page. The browser will allow the user to select a file from their computer and then save a reference to that file in the browser for later use.

When the user presses the **Send** button, we want to get a reference to the file that the user has selected. To do this, you need to add an event listener, as shown in the following code:

```
sendButton.addEventListener("click", function (event) {
  var files = document.querySelector('#files').files;

  if (files.length > 0) {
    dataChannelSend({
      type: "start",
      data: files[0]
    });

    sendFile(files[0]);
  }
});
```

You might be surprised at how simple the code is to get this far! This is the amazing thing about working within a browser. Much of the hard work has already been done for you.

Here, we will get a reference to our input element and the files that it has selected. The input element supports both multiple and single selection of files, but in this example we will only work with one file at a time. We then make sure we have a file to work with, tell the other user that we want to start sending data, and then call our sendFile function, which we will implement later in the chapter.

Now, you might think that the object we get back will be in the form of the entire data inside of our file. What we actually get back from the input element is an object representing metadata about the file itself. Let's take a look at this metadata:

```
{
  lastModified: 1364868324000,
  lastModifiedDate: "2013-04-02T02:05:24.000Z",
  name: "example.gif",
  size: 1745559,
  type: "image/gif"
}
```

This will give us the information we need to tell the other user that we want to start sending a file with the example.gif name. It will also give a few other important details, such as the type of file we are sending and when it has been modified. The next step is to read the file's data and send it through the data channel. This is no easy task, however, and we will require some special logic to do so.

Breaking down a file into chunks

Now that we have a file, we need to get it ready to send to the other user. The easiest approach to this problem is to take the entire file and call our send method with the data. This will take the entire file and give it to the data channel to process as one large message:

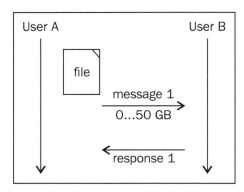

Once this happens, the data channel will take the message and try to send all the data through the connection between the clients. Let us say that the user is trying to share their entire music collection, which comes to a size of **50 GB**. The data channel will tell the SCTP protocol that it has one message that starts at **0** bytes and ends at our file size, which is **50 GB**.

This is where the problems start to arise. If the user is on a bad network connection, the message could fail to send. If we have our data channel setup in a reliable or retry mode, the underlying protocol will simply discard the currently sent data and retry to send the entire message again. This means that both users have to have a solid connection for the hours of time that it would take to send the entire file!

To alleviate the issue, we need to introduce the idea of *chunking* our file. If you have ever used an application such as **BitTorrent**, you should have an idea of how this works already. A network such as BitTorrent will split up large files into hundreds or even thousands of smaller pieces so that it is easy to transport between users. This is also similar to the concept of shipping containers. The idea is that if you break things down into smaller, same-sized pieces, you can optimize the transportation of the smaller *chunks* more easily than trying to deal with a single large one.

We can apply this concept to our file just as easily. Consider a file to be one giant stream of 0s and 1s all in a row. This is known as our binary file data. The easiest way to chunk this file is to read a few lines of binary, send it over, then read the next few lines of binary and send this over. We can then keep doing this all the way until the end of the file. If we lose connection, we can just start over at the last chunk we sent, meaning the entire progress is not lost.

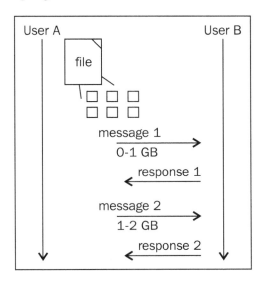

This is one of the drawbacks of working with such a low-level transport system. A lot of the work is not done for you and logic will have to be put in to handle your specific use case. It is built as a general solution for most problems, so there is a lot of work to be done to make use of the data channel.

Making chunks readable

Another topic that we have not touched on is exactly how to send binary data across the data channel. In our previous example, all of our data was already in a string format. This is due to the fact that it was in the form of simple text messages. This is a different story in the case of the File API, however. Files, such as images, audio, and more are not easily converted into human-readable strings, so we will have to work with sending binary data to the other user.

Now, you might ask why you cannot just send over the data you get directly from the File API? This would make a lot of sense, since both of the clients technically use the same language. The data channel should support this language too, right?

Turns out that life is never so easy! The unfortunate part is that there are many layers between JavaScript and the network protocol stack that handles the sending of data between clients. This stack speaks in an extremely specific and specialized way that requires you to translate the data into a more friendly and readable format.

To handle the translation, we are going to utilize **Base64 encoding**. Base64 is a binary-to-string encoding format that makes binary data easier to transport. It takes binary data and translates it into the ASCII format, which is easier for JavaScript and the network stack to handle across different platforms. This is because the protocol is well-defined and understood by many different devices. The way encoding works is that we will have to take our file data and encode it into Base64, and then decode the same data on the other side into the format we want.

To get our code started, we will define a few simple methods that will encode and decode our data. You can define these anywhere in your script so that we can use them throughout the application that we are going to build:

```
function arrayBufferToBase64(buffer) {
    var binary = '';
    var bytes = new Uint8Array( buffer );
    var len = bytes.byteLength;
    for (var i = 0; i < len; i++) {
        binary += String.fromCharCode( bytes[ i ] );
    }
    return btoa(binary);
}
```

This function will take in `ArrayBuffer`, which is what our File API is going to give us back when it reads the file's data. It allocates a new array, loops through the data and converts them to characters, then translates the binary data using the built-in `btoa` function in the browser. Now that we can encode data from one user, we need a function to decode data on the other side:

```
function base64ToBlob(b64Data, contentType) {
    contentType = contentType || '';

    var byteArrays = [], byteNumbers, slice;

    for (var i = 0; i < b64Data.length; i++) {
        slice = b64Data[i];
```

```
            byteNumbers = new Array(slice.length);
            for (var n = 0; n < slice.length; n++) {
                byteNumbers[n] = slice.charCodeAt(n);
            }

            var byteArray = new Uint8Array(byteNumbers);

            byteArrays.push(byteArray);
        }

        var blob = new Blob(byteArrays, {type: contentType});
        return blob;
    }
```

This one is a little more complex and we will not go into the finer details of how it works. The first thing it does is the opposite of the encode function. It takes the array and figures out the character code for each character sent over in the array. Once it has the resulting array, it needs to turn it into `blob`. This is so that JavaScript can easily interact with the data and is able to save the file, which we will do later on in the chapter.

Reading and sending the file

The next step is to actually read data from the file and send it to the other user. We will combine the data channel with the Base64 encoding to efficiently send the file. Let us implement the `sendFile` function that we used earlier in *Getting a reference to a file* subsection under *Getting a file with the File API* section in this chapter when getting a reference to the file:

```
var CHUNK_MAX = 16000;
function sendFile(file) {
  var reader = new FileReader();

  reader.onloadend = function(evt) {
    if (evt.target.readyState == FileReader.DONE) {
      var buffer = reader.result,
          start = 0,
          end = 0,
          last = false;

      function sendChunk() {
        end = start + CHUNK_MAX;
```

```
        if (end > file.size) {
          end = file.size;
          last = true;
        }

        dataChannel.send(arrayBufferToBase64(buffer.slice(start,
end)));

        // If this is the last chunk send our end message,
otherwise keep sending
        if (last === true) {
          dataChannelSend({
            type: "end"
          });
        } else {
          start = end;
          // Throttle the sending to avoid flooding
          setTimeout(function () {
            sendChunk();
          }, 100);
        }
      }

    sendChunk();
    }
  };

  reader.readAsArrayBuffer(file);
}
```

The first thing you will notice is that we instance a new object that we have not seen earlier we call `FileReader`. This is an object that comes from the File API. It encapsulates a few different methods to read files in JavaScript using different formats. In this example, we read the file as `ArrayBuffer`. This is typically used when reading lower-level binary data from files.

Once the file has completely been read, we go through a series of steps:

1. Checking and making sure `FileReader` is in the `DONE` state.

2. Setting up our initial variables and getting a reference for the buffer of file data.

3. Creating a recursive function that we can send pieces of the file with.

4. In our function, we start from `0` and read a set of bytes that are equal to one chunk.

5. We then make sure we are not past the end of the file since there is nothing to read past the end.

6. The data is encoded and sent in the Base64 format using the functions we set up earlier.

7. If this is the last chunk, we tell the other user that we have completed sending the file.

8. If we still have data to send, we send the next chunk after a certain amount of time to avoid flooding the API.

9. Finally, we start the recursive process by calling `sendChunk`.

At this point, you can test out the sending of files using the API. You should be able to connect two users, select a file, and see the chunks being received on the other side. Go ahead and play with the example to see what happens when you send smaller files as well as larger ones. You should also try different types of files to see what happens when you send a PNG versus a TXT file and so on.

Putting it together on the other side

Now the other user should have a collection of chunks that make up the entire file. Since we used the `ordered` option in the data channel, the chunks should have come from the other user in order. This means we can put them together in one large data object as we receive them from the other side.

To store our pieces of data, we can push them into a simple array data structure in the client. We will do this directly when the data channel reports that a message has been received. We will also use the opposite decoding function that we covered in the *Making chunks readable* subsection under the *Breaking down a file into chunks* section in this chapter, to put our chunks into the right format:

```
// Add some new global variables
var currentFile = [],
    currentFileMeta;

// Add this to the openDataChannel function
dataChannel.onmessage = function (event) {
    try {
      var message = JSON.parse(event.data);

      switch (message.type) {
        case "start":
          currentFile = [];
          currentFileMeta = message.data;
          console.log("Receiving file", currentFileMeta);
          break;
        case "end":
          saveFile(currentFileMeta, currentFile);
          break;
      }
    } catch (e) {
      // Assume this is file content
      currentFile.push(atob(event.data));
    }
};
```

You can see we added two new global variables to hold the current file data that is being sent, as well as the metadata telling us about the file. Now we can fill out the `onmessage` handler inside the `openDataChannel` function. Initially, we will try to use the JSON parser to decode the contents of the message. The `JSON.parse` API will throw an error if it is unable to decode the data, meaning that it is not in JSON format. This is what we will use to inform us if the message is a file data or a command coming from the other user.

The two JSON commands that we have are the `start` and `end` triggers. The `start` trigger tells us to set up our local variables to accept new file data and gives us the file's information. The `end` trigger tells us that the file transfer is complete and we can now save the file to the user's hard drive in some manner.

If the message is not in JSON format, we will assume that this is a chunk of file data. The data will get decoded and inserted into our list of data. This will contain all of the chunks that we get until we save the file later on.

Once we finally get the end event, we need to save the file to the user's system. One big drawback of the browser is that it is a **sandbox environment**. What this means is that nothing that happens inside the browser should be able to effect the user's OS in any way. This is to prevent a malicious website from deleting or modifying files on the user's system in some way. Instead, we will use a safe method to save the file to the system.

Our approach to saving the file is similar to the way the user would download a file from a website. Typically, a website will host a file on a server somewhere and provide a URL to the file. When the browser recognizes that the link is directly linked to a file, it will prompt the user on where they would like to save it or directly view it in the browser, depending on the type. Using a similar strategy, we can add the `saveFile` function to our project:

```
function saveFile(meta, data) {
    var blob = base64ToBlob(data, meta.type);

    var link = document.createElement('a');
    link.href = window.URL.createObjectURL(blob);
    link.download = meta.name;
    link.click();
}
```

The first thing our function does is take our file data and converts it to `blob`. This is done using the same Base64 encoding that we covered in *Making chunks readable* subsection under *Breaking down a file into chunks* section in this chapter. Next, we can create a new `link` element that we can simulate when the user clicks on it. We can set this link's location to a fake URL that points to our data.

We create this fake location using another browser API called `createObjectURL`. This is a specialized function that creates locations for file and blob-based JavaScript objects. Once we have this fake location, we can assign it to our link so that when a user clicks on it, they are prompted to navigate to the location. Since our location points to a file, it will ask the user if they would like to download the file.

Showing progress to the user

A final touch that we can add to improve our application is a progress meter to show how much of the file the user has and how much they have left. This makes for a better user experience so they can see when something is happening. We need to tell the first user how much data we have sent and the second user how much data has been received.

To show how much we have sent, we can add the following code to the `sendChunk` function:

```
if (end > file.size) {
    end = file.size;
    last = true;
} // Code that already exists

var percentage = Math.floor((end / file.size) * 100);
statusText.innerHTML = "Sending... " + percentage + "%";
```

We just take the current number of bytes sent versus the file's size and convert this into a percentage. Now, we can do the same for the receiving user:

```
currentFile.push(atob(event.data)); // Code that already exists

currentFileSize += currentFile[currentFile.length - 1].length;

var percentage = Math.floor((currentFileSize /
currentFileMeta.size) * 100);
statusText.innerHTML = "Receiving... " + percentage + "%";
```

We can apply the same logic on the other side, this time getting the current size from our last chunk's size. We then convert this to a percentage and show it to the user. You should now have a simple but effective file sharing application as shown here:

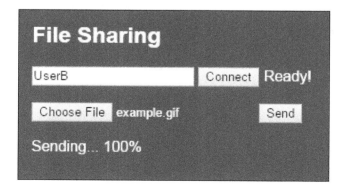

Both users should now be able to see what percentage of the file has been sent between the users. This is a great thing to experiment with on different computers over different networks to see just how well the data channel performs. Also, try different sizes of file and see what kind of results you get.

Self-test questions

Q1. The latest version of the File API has been improved in HTML5 to provide much more functionality than before. True or False?

Q2. Breaking down a file into chunks assists in:

1. Sending over slow networks
2. Recovering chunks that get lost in transportation
3. Sending really large files
4. All of the above

Q3. Since both computers speak in binary language, it is easy to send the binary data from a file over the data channel. True or False?

Q4. A file is saved in our example much the same way as:

1. Saving a file in Microsoft Word
2. Downloading a file from a link
3. Copying and pasting a file

Q5. The File API runs inside a sandbox environment, which means that it is prevented from working like other native applications. True or False?

Summary

It is pretty amazing what we were able to accomplish in just one small chapter. The example that we built works much like any file sharing application that is out there today. We were able to easily combine APIs together to create a fully featured experience in just a few hundred lines of code. It even works efficiently and accounts for several things, such as slow networks and large files.

This is truly an example that you can expand upon and experiment with. There are plenty of ways to make this application much better to use. The first step could be printing out more information for the user to see, including what file they are sending and how large the file is. You can also give more information about the progress of the file and even the download speed. The user could also be given the chance to download the file whenever they want instead of immediately downloading the file right away, or even given the ability to input a file name to download it as.

The other drawback of our application is the size of a file. Right now, the entire file gets loaded into memory and saved as one huge download. This makes it hard to transfer larger files that could consume more space than the amount of memory on the computer. If you are really looking for a challenge, see how you can utilize the File API to reach the limits of file transfer, transferring files that are as large as 10 GB or 15 GB.

In the final pages of the book, we are going to cover some more conceptually advanced techniques when using the data channel. These topics are a great read if you are looking to start using the data channel in a production-ready environment. We will start diving deep into the inner workings of how large-scale WebRTC applications are already working today and the battles that they have had to face along the way.

8
Advanced Security and Large-scale Optimization

Up to this point, we have covered a shallow portion of the entirety of WebRTC. This has been limited to what we could build in a chapter inside our local computer without hooking up to any real services. This is great if it's only limited to you or a few of your friends but, unfortunately, this will not lead you to being able to connect thousands of people across the world. To do this, we will have to dive into deeper and more advanced topics, such as security, performance, and supporting large-scale networks.

The aim of this chapter is to provide a small amount of information on these topics so that you can research them in more depth at your leisure. Most of the information will be conceptual and we will not build a working demo in this chapter. By the end of the chapter, you should have a good idea of what concerns you will have when building a large-scale WebRTC-based service and where to go for more information.

Researching and reading is the key to staying up-to-date with the latest technology and techniques. As you read through this chapter, feel free to stop and research each topic on the Web. The goal is to spark ideas when implementing your own WebRTC application.

Securing the signaling server

What we have built so far has been in the spirit of teaching a single part of the WebRTC API. This means many shortcuts were taken and we sacrificed performance and security to make it easier to learn. Our signaling server, although it works, is simple in nature and can be upgraded to support a much larger set of features.

Using encryption

The largest and most obvious upgrade is mandatory encryption of the signaling server. Encrypting the messages of the signaling server will ensure that no one can intercept a message to the server, thus figuring out which clients are talking to whom. This is easily the largest gap in security that the signaling server we built has right now. It is also the easiest to patch up since encryption is a highly standardized and widely used technology on the Web today.

The two standard encryption methods for our signaling server are HTTPS and WSS. You should recognize HTTPS by now as the standard SSL encryption for websites over HTTP. This should be fairly easy to research and implement with the resources already available on the Web today. WSS is a WebSocket-based SSL security over the TLS protocol. It works in a similar way to how standard HTTPS works when the correct certificates are set up on your web server. It is important that you research these two and fully understand them when moving your WebRTC application into the real world!

Using an OAuth provider

One of the other major upgrades to our example would be the integration of an identity provider. An identity provider is software that allows your application to keep track of users using another application's identity data. One good example of this is the Google login. This would provide each user with a unique user ID as well as their list of contacts to call. It would also add a level of security to our signaling server so that only two users who know each other can start a call.

Today's world of the open Web gives many choices when authenticating users. One of the most popular authentication technologies at the time of writing this is **OAuth**. OAuth is a set of open standards that allows users to authorize access to data belonging to them from other services. In our case, we want our users to give us a unique identification token along with a list of contacts that they would like to talk to. This will also add to the security of our server by making sure only authenticated users are allowed to access our application:

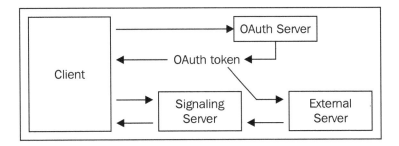

The first step on this path is figuring out how and where to add the identity provider to secure your server. In previous chapters, we showed you a diagram of how the signaling server and the WebRTC connection interact with each other in order to create a connection between two users. We have a few concerns when implementing this:

- Both users should have to log in before they can make any calls
- Both users should know about each other before they are allowed to make any calls
- Neither of the users should be able to be spoofed by an online attacker

To ensure that both users authenticate themselves before they make any calls, the authentication mechanism should be in place before the user can interact with the signaling server. If the user has to log in in order to do anything, this will reduce the exposed servers that a user can maliciously try to hack into.

Once you choose an OAuth service, you can begin integrating it with your signaling server. The first thing the user should see is the opportunity to log in to your service either through a third party or your own authentication service. The client should then give your WebRTC application access to their data using OAuth technology.

The way it works behind the scenes is through the use of tokens. A token is a random string of numbers and letters that represents a temporary access key to a set of data. When the client authenticates via OAuth, the JavaScript code will receive the token generated by the third party. The client code can then send this code to your signaling server, which can in turn verify and access the user's data from the third party.

One analogy used by the OAuth getting started guide is that OAuth is a **valet key** for web resources. The key gives you temporary access to a limited set of resources on another server.

Each third-party service has their own getting started guide and tutorials on how to get set up with access to user data from their platform. Keep in mind, third party can also mean your own company's authentication service as well. Company A can separate their OAuth login service from their own products and applications, giving their software better scalability and maintainability. We will not go into much detail here and leave it as an exercise for you to research more on how OAuth works.

Using encryption and integrating a basic OAuth service should give your project the simplest of security features to ensure only verified clients get connected in your application. However, this is only the tip of the iceberg when it comes to user security and authentication. The one thing that is certain is that security is extremely important to any application, especially ones based on WebRTC.

Supporting mobile devices

With the meteoric rise of the Web on mobile devices, most developers will have to support WebRTC calling on phones at some point. It is the medium that is most often paired with the idea of web-based calling. The idea of ditching expensive minute plans and sticking with a data service is an alluring concept. Also, a web page can make use of the microphone and camera already attached to the device instead of needing to purchase these items for a desktop computer. In these cases, however, the low power and cellular-based network connectivity can become an issue for streaming applications.

While sending data between a desktop and a mobile phone, the first major issue will be the data connection. As fast as phone networks are getting, the connection speed will still be much lower than a standard desktop computer wired into a wall. If the phone does not have a great connection, the call quality of a WebRTC connection will be much worse than a standard phone call that users are used to.

One of the most often-looked-over issues for phones is connectivity. Let us take a moment to compare a WebRTC calling service to your typical cell phone plan. Even if you have 1 million users that make just one call per day, even successfully connecting 99 percent of these users still results in 10,000 missed calls. This fact is complicated by the number of different devices to support, as well as the multiple servers and steps required to actually make a call. Not to mention the fact that phone companies are required to provide emergency service connectivity and fallback plans for multiple networks! A tough problem to solve, indeed.

To solve the data connection issue we need to simply reduce the amount of data that we send over the network. The data that we send is measured by the number and size of each frame of video that is captured. These are packaged and encrypted, then sent across the user's mobile network frame-by-frame to the other side. To reduce this, we want to reduce the size of the video that the users are capturing based on the device. Here is the code required to do this in our application:

```
var mobile = {
  video: {
    mandatory: {
      maxWidth: 640,
      maxHeight: 360
    }
  }
};

var desktop = {
  video: {
    mandatory: {
      minWidth: 1280,
      minHeight: 720
    }
  }
};

var constraints;

if (/Android|webOS|iPhone|iPad|iPod|BlackBerry|IEMobile|Opera
Mini/i.test(navigator.userAgent)) {
  constraints = mobile;
} else {
  constraints = desktop;
}

navigator.getUserMedia(constraints, success, function (error) {
  // Possibly try again at a different resolution
});
```

We need to define a set of constraints for each type of device. In this case, we only define one for mobiles and the other for desktops. We then test against the user agent string to see if this is potentially a mobile device and decide what constraints are used. This is passed into the getUserMedia call where the browser will try to figure out if the device can support these constraints. If not, it will call the error callback in which you can try a different set of constraints to see if these will work.

You might have noticed that if a mobile device is talking to a desktop computer, it will send a smaller resolution stream but still receive a higher resolution one. This means that we can reduce the size of our data transport more if the client is mindful of what device the other user is operating on. We can do this by adding more commands to our signaling server in order to exchange the device type before connecting.

This should at least be a good starting point to support basic connectivity between mobile devices. There are many more optimizations we can do around reconnecting if the network goes down, sending even less data, and changing the resolution while the call is running. The optimizations you will need to implement will change based on the needs of your application. Be sure to list what kind of connectivity your WebRTC application will need on mobiles and test on many mobile devices to solve problems.

Introduction to mesh networking

Once your WebRTC application is working on multiple devices with a reasonable amount of security, the next question is: how do I scale this up to multiple users? One-to-one calls are great, but what if we can connect several users together inside one call? This is where we begin to dive into the world of **mesh networking**.

Mesh networking is a topic that has been around for a while. It's a term used in almost any technology that deals in the networking of multiple computers. In its most basic sense, a mesh network represents a set of computers that can all communicate with each other directly. Each **node** is connected with every other **node** in the network and no one **node** is responsible for the entire network:

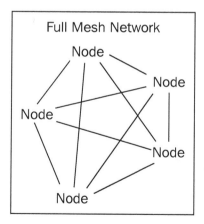

There is a good chance that you have encountered many mesh networks in your lifetime. For example, the Internet is already a form of mesh network. So, when you send out a request for a certain web page, every **node** on the network works together to deliver that page to your computer. Mesh networks can have every **node** talk to each other in a full mesh, or have only a subset of **nodes** connect together in a partial mesh.

You can already see how this type of concept can play into a peer-to-peer technology such as WebRTC. Currently, every call is a mesh network of users where all the nodes in our system are connected to each other. Each node is a user and we have two **nodes** in every call that we set up.

Now we can expand this idea and add more **nodes** to our network. If two users can connect together then why not a third? There is no limit on the number of WebRTC connections you can have open in one browser at a time. Each user in our network can connect to multiple people at the same time, thus creating a mesh of users all in one call:

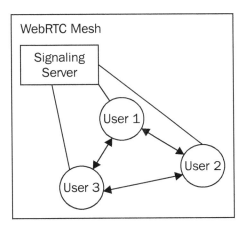

In the preceding diagram each one of our users would begin by connecting to the **signaling server**. They would then request calls to all the users they want to communicate with. Each user would go through the request/response phase with every other user they are connecting to. This would set up three WebRTC connections, allowing everyone to communicate with each other at the same time.

Even though we can imagine making a 100 person call, that does not mean that the technology will scale that far. There is a cost associated with each connection and the number of connections will multiply with each user added. Later in the chapter, we will talk about scaling beyond the limits of a simple mesh.

Types of network meshes

Now that we have figured out what mesh networks are, we can talk about the different types of mesh networks that we can implement. As with previous sections, the first thing we want to do is define a set of constraints to work within and then explain how each mesh type is better or worse for each constraint. The main goal here is to connect the most users possible in the most efficient way so there is a good experience for all our users.

Here is a set of simple criteria to judge each mesh type by:

- What is the minimum or maximum bandwidth a specific user might have?
- What is the maximum number of users I want to connect?
- How much data loss is acceptable for any given user?

Everyone to everyone

The first type of network we will cover is one where every node connects to every other node. This network, by far, is the easiest to implement since the logic is simple. The server just has to keep track of everyone that joins the call and each person should broker a WebRTC connection with every other user in the list.

Even though this is easy to implement, it comes with many drawbacks. The bandwidth of this network goes up significantly with each user that connects. Each user in the network has to send their stream data to every other user in the network. This means that the saturation point of the call would be reached after only a few users join. This gives us a low maximum number of users, and the potential for data loss occurs when too many users join the call. This strategy is great when making a simple three or four person call but not much more than this.

Star network

To support more users, we have to be smarter about the way we connect them. One simple way we can do this is to reduce the number of connections we have within the call. We can use a star topology to alleviate the issues we have with connecting everyone on the call. The star network works by having everyone connect to one **host** node, and then this node is responsible for sending the correct video streams to all the other nodes on the network:

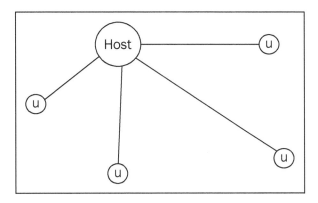

The **host** node in our case can be any one of the users on the call. All the other users would make a WebRTC connection to the **host** user and that user would then relay the streams to all of the other parties. Since WebRTC is agnostic among browsers, the code would be able to take the stream from one connection and add it to another. This would mean that the network would have less connections overall since there would be a number of connections equal to the number of callers minus one. On the other hand, we still transfer a similar amount of data between our users, so although we may get more connections and less dropped data than a complete mesh, it will not be by much.

 This is another unfortunate moment where I have to stop and talk about browser support. Unfortunately, the ability to add multiple streams to one WebRTC connection is still an in-progress item for some browsers. The streaming technology is vast and complex, so it may be some time before star networks using clients will be viable. This being said, we will talk about the **Multipoint Control Unit** (**MCU**), which helps solve this issue, later on in the chapter.

One issue that we will have to solve is the ability to actually select a user to be the host on the call. This is a feature that could be added to the signaling server. When the user logs in, they should collect a certain amount of information about the user and send it to the signaling server. At this point, when the user initiates a call, the signaling server can compare the data for each user and make a selection based on a set of criteria:

- The device type helps select users based on processing power and bandwidth. Desktop computers should be prioritized over mobile users.

- The video encoding support could give a boost if one user can encode and decode video in more or faster formats than other users.

- We can test the user's bandwidth using a small file download to see what type of network the user is on.

The last piece of criteria in this case can be the most helpful as it directly relates to the environment the user is on. A user may be using a laptop computer, which has more power than a mobile device but if they are tethered to a mobile phone, it will not be faster than a mobile phone connected to a high-powered Wi-Fi router. The strategy to do this is to make an image download of a known size, and time how long it takes to download this file:

```
var src = "example-image.jpg",
    size = 5000000,
    image = new Image(),
    startTime,
    endTime,
    totalTime = 0,
    speed = 0;

image.onload = function () {
  endTime = (new Date()).getTime();
  totalTime = (endTime - startTime) / 1000;
  speed = (size * 8 / totalTime); // bytes per second
};

startTime = (new Date()).getTime();
image.src = src + "?cacheBust=" + startTime;
```

The logic here is fairly straightforward:

- We set up a new image along with defining the known size of the image beforehand
- Set up a load handler to determine the speed of the download once the image is loaded
- Set `startTime` of the download as the current date/time in JavaScript
- Set the source of the image that begins the download

This will give us a good idea of how strong or fast the network connection is for the user. It might be useful to do this a few times as the network strength can vary over time. Once we have collected a number of data points about the environment the application is running in, we can send all of this to our server. Our signaling server can then use this information to determine the best user to host the call.

Partial mesh

The last potential mesh we will cover is a partially-meshed network. A partial mesh is great in applications outside the realm of multiuser communication. This is especially helpful when using the data channel, as every node in the mesh network does not always have to be connected to every other node:

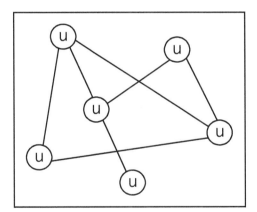

One good example that we referred to earlier is a torrent-based file sharing network. When you are sharing a file, you just need a number of chunks of data, not bothering about which node you get them from. In this case, we can select a subset of our potential users to connect to and attempt to connect to each of them in turn. These other nodes in our network might get their data from other users in the network, different from the first node's set. This means we can organize and sort our users in our network based on how many connections they currently have and the bandwidth of each user.

The basic strategy behind a partial mesh is to get the most optimal number of connections for each user. For instance, a partial mesh might have a mobile user only connect to one or two other nodes, while a desktop can connect to ten or fifteen other nodes. We can profile and sort our users based on their device information in order to give the optimal transferring of data.

Overall, the partial mesh is a powerful tool when done properly. With enough users on the network, you can have almost 100 percent uptime as well as minimal data loss. If a node disconnects, you simply connect to another node in the network and request the same data. However, the power of this network is directly proportional to how similar the data is across each node and how many users are on the network. The best thing to do is to set up a server that can sort and assign nodes to connections and then tweak the numbers and sorting algorithm until the network feels optimal to each user.

Limits of mesh networking

Although mesh networking is an easy path to enabling multiuser communication, it does have its drawbacks. In practice, mesh networking is a straightforward technique that is simple enough to understand. Implementing this technique in a production-ready environment is also fairly simple, since much of the logic can be implemented on the clients themselves. There is no need to add additional high-capacity servers other than the ones that select which nodes connect to each other.

The first drawback of the model that comes into play is how the network reacts when a single user drops out of the call. If you have ever been on a call with many people at once, you will know how difficult it can be to keep everyone connected. If one person is hosting the call and they drop out, the other users will take some time before they can recover from the loss. In a star network, this is especially bad if the host is the one who drops off of the call. Not only do we have to select a new host node but also establish multiple new connections between the other users.

Another major drawback is the support for multiple streaming rates and environments. Imagine a case where each person in the call has a vastly different network and client environment. This could include a desktop computer, mobile phone, tablet, and someone using the free Wi-Fi at a coffee shop at the same time. It will be hard for a network mesh to satisfy all clients equally and give everyone a decent user experience. In the end, someone will have to suffer.

The issue with mesh networking is that we are trading the ease of development and low cost for the user experience. A great example of this transition has been in video games over the years. Initially, many games used the star network model for the purpose of networking their users. One user would create the game and others would join the host's game, essentially creating the same pattern we reviewed earlier. Over time, as multiplayer games grew increasingly popular, the flaws of user experience were extremely visible. When a host dropped out, the entire game had to be transferred to another user, causing delays. Not only this but the host user could often cheat and edit the data of the game, making them prone to cheating.

Today, many of these video games have transferred to a server-client model of networking. With the costs of hardware and high-powered servers that can process hundreds of users at a time lowering, this model became more affordable. A server could essentially be the host in the star network, allowing the game to never get disconnected. These servers are also owned by developers, which greatly lowered the chance for users to cheat at the game and modify any data without the developers' intent. It turns out that this same theory exists in the communication world and is often called a MCU.

Video conferencing with more users

Many large-scale communication companies have made the transition to server-client methods of networking vast number of users. There have been a wide number of different solutions that work in many ways, but they are all built on the principal of using servers as nodes in the network instead of clients. These MCUs give networks better stability, performance, and a better overall user experience at the cost of being expensive.

The type of network that uses a **MCU** works similar to a star network mesh, where a server is the host machine instead of any single user. This allows the call to be controlled from a central location that has high availability and stability. It also allows the developers of the application to scale the bandwidth of the network, keeping in mind the needs of the user base:

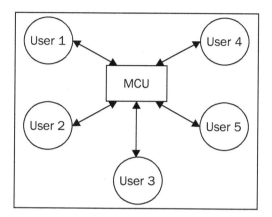

This also gives the network a lot of control over how the users connect and interact with each other. This is a fine-grain control of the user experience. The following explanation illustrates the working of a typical MCU connection:

- All users successfully connect to the MCU server hosted by the application developer
- All users on the call will send their video and audio streams to the server
- The MCU will decode and process each incoming stream individually
- The MCU has the option of taking each individual stream and re-encoding them as one single stream to save on bandwidth
- The MCU will send back an individual stream to each user, of the other users on the call

You may notice the major difference here is that the server can stitch together the other streams and only provide one stream back to each user. This is an extremely powerful feature since it cuts down the bandwidth needed to support many users in one call.

Instead of needing to send multiple streams to each user, they only have to support the bandwidth for one video stream to the server and one back. This greatly improves the user experience overall, and allows your network to scale out much larger than with a typical mesh network.

We can take this concept even further when you have full control over the network streams. If you have used something such as Google Hangouts, you may notice that there is one stream, which is currently presented while the other streams are shown as smaller boxes at the bottom of the screen. This allows the network to switch speeds on a per-user basis, giving the highest bandwidth to the user that is currently active on each machine. This is just an example of one advanced tactic that can be employed when using a MCU-like server.

For all the benefits mentioned here, you might wonder how you can actually start implementing an MCU infrastructure in production. This is the major trade-off when using this technology — it is expensive and hard to set up. As of writing this, there are a few open source MCU servers dedicated to WebRTC development, but the installation and compilation of these services are only for the most advanced users.

The other option is to go with **scaling as a service**. Companies will offer scaling WebRTC applications as a service at a premium to the developer. The company will provide all the hosting and technology preinstalled on their servers. All the developer has to do is pay the cost and follow the tutorial to set up the application with their servers. This can be a somewhat easier alternative to setting up the infrastructure yourself.

It is highly recommended that you do a lot of research into different server-based scaling techniques. The industry is a vast area of different technologies, both open and closed source, so there is much to sift through. Be prepared to spend a lot of time and money when going down this route. On the other hand, if your application requires an optimal user experience, using server-based architecture will give this to you.

The future of conference calls

This is just the tip of the iceberg when it comes to large-scale WebRTC applications. With the globalization of technology communication, platforms are not going to go out of style anytime soon. It is an industry that will continue to be invested in by many large corporations. This means that the technology presented here is going to become more accessible and powerful over time.

The big thing to remember is that none of the technology presented in this chapter is anything new for WebRTC. All of these concepts have been around for years before WebRTC hit the scene. This means there is an entire wealth of technologies and products that have already been tested and just need to be converted for usage within the WebRTC space.

On top of the large amount of technology that already exists, there are many ways to improve and innovate on the idea of multiuser communication. With the entire web framework at your disposal, there are new ways that this technology can be paired and combined with existing ones to give new user experiences.

Self-test questions

Q1. The technologies you can use to secure your signaling server include:

1. HTTPS
2. WSS
3. OAuth
4. All of the above

Q2. The biggest problem when supporting mobile users is the resolution of the camera on their mobile device. True or False?

Q3. What is a large concern when building a mesh network?

1. The OS of the user's computer
2. The browser that each user is running on
3. The bandwidth available on the user's network
4. What OAuth method the user logged in with

Q4. A star network is where everyone connects to one host node in the network and this node serves all the other users. True or False?

Q5. Mesh networks give the most optimal user experience in favor of ease of development and low cost to the developer. True or False?

Q6. An MCU-based network is most comparable with a:

1. Torrent network
2. Star network mesh
3. Full network mesh
4. Partial mesh network

Summary

By now, your head should be filled with dreams of the amazing things that WebRTC can do. We covered a large list of advanced techniques that can be used to greatly enhance the usability and performance of your application. Each of these topics has a wealth of information to read about and also discover on your own. It would be a great idea to spend time researching each one at your leisure if you plan on continuing your WebRTC education!

All these topics are aimed at learning how to release a large-scale WebRTC application. When learning how to use WebRTC, developers may not think about what happens when you go from two users to two hundred or two thousand users at a time. There can be a lot of growing pains if you are not prepared for what may happen when adding more users to your application. The best way to learn is to keep experimenting with the techniques presented here and pick and choose what your most immediate needs are.

The examples that we built upon are a great way to jump into the WebRTC space. Everything that was presented here is easily built upon and can be refactored into something much greater than we built over the course of these chapters. This, in a sense, is the beauty of using web technologies and, more specifically, WebRTC. Using web standards opens the doors to integrating not only with more libraries and frameworks but also with third parties and applications out there. WebRTC not only involves connecting users through communication but also connecting applications by providing more features experienced across platforms. It will be an exciting technology to keep an eye on in the years to come, to see how it changes people's thoughts about web development.

Answers to Self-test Questions

This appendix contains answers to all the self-test questions that appear at the end of every chapter. Now, let's have a look at the answers to respective questions.

Chapter 1, Getting Started with WebRTC

Q1	True
Q2	3
Q3	False
Q4	4

Chapter 2, Getting the User's Media

Q1	False
Q2	3
Q3	True
Q4	1
Q5	True

Chapter 3, Creating a Basic WebRTC Application

Q1	True
Q2	False
Q3	3
Q4	True
Q5	4

Chapter 4, Creating a Signaling Server

Q1	True
Q2	2
Q3	4
Q4	True

Chapter 5, Connecting Clients Together

Q1	False
Q2	4
Q3	False
Q4	3

Chapter 6, Sending Data with WebRTC

Q1	False
Q2	1
Q3	True
Q4	False
Q5	4

Chapter 7, File Sharing

Q1	True
Q2	4
Q3	False
Q4	2
Q5	True

Chapter 8, Advanced Security and Large-scale Optimization

Q1	4
Q2	False
Q3	3
Q4	True
Q5	True
Q6	2

Index

A

audio
  communicating with 1, 2
  enabling, on Web 2
audio feedback 16

B

Base64 encoding 130
BitTorrent 128
browser supports
  URL 6

C

call, client application
  hanging up 94
  initiating 91, 92
  traffic, inspecting 92, 93
call, signaling server
  answering 69, 70
  hanging up 71
  ICE candidates, handling 70
  initiating 68
Canvas API 24
chunks
  collecting 133, 134
  file, breaking down into 128, 129
  making readable 129-131
client application
  about 82, 83
  call, hanging up 94
  call, initiating 91, 92
  connection, getting 85, 86
  HTML page, setting up 83, 84
  improving 100

  logging in to 87, 88
  peer connection, starting 88, 90
  traffic, inspecting 92, 93
  WebRTC client 94
codec 3
conference calls
  future 153, 154
constraints
  URL 17
Content Delivery Network (CDN) 114

D

data channel, RTCDataChannel object
  about 104, 107
  ArrayBuffer type 108
  ArrayBufferView type 108
  Blob type 108
  id option 107
  maxRetransmits option 107
  maxRetransmitTime option 107
  negotiated option 107
  ordered option 107
  protocol option 107
  reliable option 107
**Datagram Transport Layer
    Security (DTLS)**
  about 104,109
  specification, URL 110

E

encryption 109, 110
everyone to everyone network
  about 146
  drawbacks 146

F

file
  breaking down, into chunks 128, 129
  obtaining, with File API 118-120
  page, setting up 121, 126
  reading 131-133
  reference, obtaining 126, 127
  sending 131-133
  sharing 117
frames per second (fps) 23

H

host node 147
host user 147

I

ICE candidates
  finding 50, 51
  handling 70
Interactive Connectivity Establishment (ICE) 42
Internet Engineering Task Force (IETC) 5

M

MCU connection
  working 152
media devices
  accessing 11, 12
  first MediaStream page, creating 13-15
  static server, setting up 12
media stream
  constraining 15, 16
  modifying 26-28
mesh networking
  about 144, 145
  limitations 150, 151
  node 144
  signaling server, connecting 145
Meteor JavaScript framework 62
mobile devices
  supporting 142, 143
multiple devices
  handling 22, 23
Multipoint Control Unit (MCU) 147, 151

N

network meshes
  everyone to everyone 146
  partial mesh 149, 150
  start network 147-149
  types 146
Node.js
  URL 12
node package manager (npm) 12

O

onclose listener 112
onerror listener 112
onmessage listener 112
onopen listener 112

P

partial mesh network 149, 150
photo booth application
  creating 24-26
progress meter
  showing, to user 136, 137

R

real-time transfer 31-34
RTCDataChannel object
  about 105-107
  data channel options 107
  data, sending 108
  states 106
RTCPeerConnection object 34, 35

S

sandbox environment 135
scaling as a service option 153
SCTP
  about 103, 104
  chunks 105
  data transportation 103, 104
  endpoints 105
  features 104
  messages 105
  specification 105

X

Thank you for buying
Learning WebRTC

About Packt Publishing

Packt, pronounced 'packed', published its first book, *Mastering phpMyAdmin for Effective MySQL Management*, in April 2004, and subsequently continued to specialize in publishing highly focused books on specific technologies and solutions.

Our books and publications share the experiences of your fellow IT professionals in adapting and customizing today's systems, applications, and frameworks. Our solution-based books give you the knowledge and power to customize the software and technologies you're using to get the job done. Packt books are more specific and less general than the IT books you have seen in the past. Our unique business model allows us to bring you more focused information, giving you more of what you need to know, and less of what you don't.

Packt is a modern yet unique publishing company that focuses on producing quality, cutting-edge books for communities of developers, administrators, and newbies alike. For more information, please visit our website at www.packtpub.com.

About Packt Open Source

In 2010, Packt launched two new brands, Packt Open Source and Packt Enterprise, in order to continue its focus on specialization. This book is part of the Packt Open Source brand, home to books published on software built around open source licenses, and offering information to anybody from advanced developers to budding web designers. The Open Source brand also runs Packt's Open Source Royalty Scheme, by which Packt gives a royalty to each open source project about whose software a book is sold.

Writing for Packt

We welcome all inquiries from people who are interested in authoring. Book proposals should be sent to author@packtpub.com. If your book idea is still at an early stage and you would like to discuss it first before writing a formal book proposal, then please contact us; one of our commissioning editors will get in touch with you.

We're not just looking for published authors; if you have strong technical skills but no writing experience, our experienced editors can help you develop a writing career, or simply get some additional reward for your expertise.

WebRTC Blueprints

ISBN: 978-1-78398-310-0   Paperback: 176 pages

Develop your very own media applications and services using WebRTC

1. Create interactive web applications using WebRTC.

2. Get introduced to advanced technologies such as WebSocket and Erlang.

3. Develop your own secure web applications and services with practical projects.

Getting Started with WebRTC

ISBN: 978-1-78216-630-6   Paperback: 114 pages

Explore WebRTC for real-time peer-to-peer communication

1. Set up video calls easily with a low bandwidth audio only option using WebRTC.

2. Extend your application using real-time text-based chat, and collaborate easily by adding real-time drag-and-drop file sharing.

3. Create your own fully working WebRTC application in minutes.

Please check **www.PacktPub.com** for information on our titles

WebRTC Cookbook

ISBN: 978-1-78328-445-0   Paperback: 230 pages

Get to grips with advanced real-time communication applications and services on WebRTC with practical, hands-on recipes

1. Explore task-based recipes on integrating your WebRTC application with systems such as Asterisk and Freeswitch.

2. Set up cutting-edge communicating networks by understanding the fundamentals of debugging, security, integration, attendant services, and more.

3. Gain advanced knowledge of WebRTC with these incredibly effective recipes.

Learning WebRTC Application Development [Video]

ISBN: 978-1-78398-990-4   Duration: 02:33 hours

Create fast and easy video chat applications the WebRTC way

1. Harness some of WebRTC's awesome features including the connection mechanisms and the resources' acquisition.

2. Set up the optimal backend to get your application up and running.

3. Discover PeerJS, the WebRTC wrapper library, for quick integration of WebRTC technology in your application.

Please check **www.PacktPub.com** for information on our titles